NATIONAL STANDARD OF THE PEOPLE'S REPUBLIC OF CHINA

Code for Installation of Intelligent Building Systems

GB 50606 – 2010

Chief Development Department: Ministry of Housing and Urban-Rural Development of the People's Republic of China

Approving Department: Ministry of Housing and Urban-Rural Development of the People's Republic of China

Implementation Date: February 1, 2011

China Architecture & Building Press

Beijing 2014

图书在版编目(CIP)数据

智能建筑工程施工规范 GB 50606-2010/中华人民共和国住房和城乡建设部组织编译. —北京：中国建筑工业出版社，2014.10
(工程建设标准英文版)
ISBN 978-7-112-16908-5

Ⅰ.①智… Ⅱ.①中… Ⅲ.①智能化建筑-工程施工-建筑规范-中国-英文 Ⅳ.①TU243-65

中国版本图书馆 CIP 数据核字(2014)第 111276 号

Chinese edition first published in the People's Republic of China in 2010
English edition first published in the People's Republic of China in 2014
by China Architecture & Building Press
No. 9 Sanlihe Road
Beijing, 100037
www. cabp. com. cn

Printed in China by BeiJing YanLinJiZhao printing CO. ,LTD

© 2010 by Ministry of Housing and Urban-Rural Development of
the People's Republic of China

All rights reserved. No part of this publication may be reproduced or transmitted in any form or
by any means, graphic, electronic, or mechanical, including photocopying, recording,
or any information storage and retrieval systems, without written permission of the publisher.

This book is sold subject to the condition that it shall not, by way of trade or otherwise, be lent,
re-sold, hired out or otherwise circulated without the publisher's prior consent in any form of
blinding or cover other than that in which this is published and without a similar condition
including this condition being imposed on the subsequent purchaser.

ISBN 978-7-112-16908-5(25704)

Announcement of Ministry of Housing and Urban-Rural Development of People's Republic of China

No. 668

Announcement of Publishing the National Standard of *Code for Installation of Intelligent Building Systems*

Code for Installation of Intelligent Building Systems has been approved as a national standard with a serial number of GB 50606-2010, and shall be implemented on February 1, 2011. In this code, Articles (Items) 4.1.1, 8.2.5(10), 9.2.1(3) and 9.3.1(2) are compulsory provisions and must be enforced strictly.

Authorized by the Research Institute of Standards and Norms of Ministry of Housing and Urban-Rural Development of the People's Republic of China, this code is published and distributed by China Planning Press.

Ministry of Housing and Urban-Rural Development of the People's Republic of China

July 15, 2010

Foreword

According to the requirements of Document JIANBIAO [2008]No. 102 issued by Ministry of Housing and Urban-Rural Development of the People's Republic of China - "Notice on Printing the Development and Revision Plan of National Engineering Construction Standards and Codes in 2008 (the First Batch)", this code was compiled by Tongzhou Construction General Contracting Group Co., Ltd. and CITIC Construction Co., Ltd. jointly with the organizations concerned.

The drafting panel examined and finalized this code by conducting extensive investigation and study and summarizing practical experiences during the preparation of this code and on basis of extensively soliciting for opinions.

This code comprises 17 chapters and 2 appendixes, and its main technical contents include: general provisions, terms, basic requirements, comprehensive pipeline, general cabling system, information network system, satellite TV reception and cable TV system, electronics conference system, broadcasting system, information facilities system, information application system, building automation system, fire alarm and control system, security system, intelligentized integration system, lightning protection and grounding and machine room engineering etc.

In this code, the provisions printed in bold type are compulsory ones and must be enforced strictly.

Ministry of Housing and Urban-Rural Development is in charge of the administration of this code and the explanation of the compulsory provisions and Tongzhou Construction General Contracting Group Co., Ltd. is responsible for the explanation of specific technical contents. All relevant organizations are kindly requested to sum up and accumulate your experience in actual practices during the process of implementing this code. The relevant opinions and advices, whenever necessary, can be posted or passed on to Tongzhou Construction General Contracting Group Co., Ltd. (No. 34, Xinjin Road, Tongzhou City, Jiangsu Province, 226300, China; Tel: 0513-86529132; Fax: 0513-86512940).

Chief development organizations, participating development organizations, chief drafting staff and chief examiners of this code:

Chief Development Organizations:
  Tongzhou Construction General Contracting Group Co., Ltd.
  CITIC Construction Co., Ltd.

Participating Development Organizations:
  Intelligent Building Branch of China Construction Industry Association
  Silian Intelligence Technology Share Co., Ltd.
  Tsinghua Tongfang Co., Ltd.
  CSCEC Electronic Engineering Co., Ltd.
  Sichuan College of Architectural Technology
  Tongzhou Construction General Construction Intelligent Communication System Engineering Co., Ltd.

Beijing Union University
Taiji Computer Corporation Limited
Nantong Huarong Construction Group
Nantong Zhuoqiang Construction Engineering Co., Ltd.
Guangzhou Fudan Auto Science & Technology Inc.
Beijing Chatone Computer Room Equipment and Engineering Co., Ltd.
Tellhow Sci-Tech Co., Ltd.
Shanghai Credy Intelligent Technology Co., Ltd.
Shenzhen Sunwin Intelligent Co., Ltd.
Xiamen Best Information Technology Co., Ltd.

Chief Drafting Staff:

Ju Qizhong Fan Tongshun Hong Bo Dong Yu'an
Xu Zhenxi Yan Lingyun Chen Xi Ding Chunying
Miao Di Chen Jiawei Yang Zhirong Guan Xiaomin
Gu Keming Kan Zhiyong Wang Yuhong Li Cuiping
Pi Youxin Zhang Xinming Li Hua Yuan Shaobin
Zhu Jingming Su Wei Zou Chaoqun Zhao Xiaobo
Li Hui Zhan Lin Ju Hongcheng Fei Xuejun
Shi Yi Zhang Qi Zhang Jinrong Zhang Bing
Cui Chunming Pang Hui Zhang Jianxin Wang Dawei
Li Jing Feng Qihua Wang Xuxin Cao Wei
Yang Xiaojun Liu Guangbo

Chief Examiners:

Xu Ronglie Xu Zhengzhong Guo Weijun Cao Yang
Chen Zhixin Chen Jianli Cheng Jun Mao Jianying
Lu Debao Ju Erlian Yang Zhushi

Contents

1 General Provisions ·· (1)
2 Terms ·· (2)
3 Basic Requirements ·· (4)
   3.1 General Requirements ·· (4)
   3.2 Construction Management ··· (4)
   3.3 Construction Preparation ·· (4)
   3.4 Project Implementation ·· (6)
   3.5 Quality Assurance ··· (6)
   3.6 Finished Product Protection ··· (7)
   3.7 Quality Record ··· (8)
   3.8 Safety, Environmental Protection and Energy Conservation Measures ········ (9)
4 Comprehensive Pipeline ·· (10)
   4.1 General Requirements ·· (10)
   4.2 Construction Preparation ·· (10)
   4.3 Pipeline Installation ··· (10)
   4.4 Wire Cable Laying ·· (13)
   4.5 Quality Control ··· (13)
   4.6 Self-examination and Test ··· (14)
5 General Cabling System ·· (15)
   5.1 Construction Preparation ·· (15)
   5.2 Wire Cable Laying and Equipment Installation ································ (15)
   5.3 Quality Control ··· (16)
   5.4 Channel Test ·· (16)
   5.5 Self-examination and Test ··· (16)
   5.6 Quality Record ··· (18)
6 Information Network System ·· (19)
   6.1 Construction Preparation ·· (19)
   6.2 Equipment and Software Installation ·· (19)
   6.3 Quality Control ··· (20)
   6.4 System Debugging ··· (20)
   6.5 Self-examination and Test ··· (23)
   6.6 Quality Record ··· (23)
7 Satellite TV Reception and Cable TV System ······································ (24)
   7.1 Construction Preparation ·· (24)
   7.2 Equipment Installation ··· (24)
   7.3 Quality Control ··· (26)
   7.4 System Debugging ··· (27)

7.5	Self-examination and Test	(27)
7.6	Quality Record	(30)

8　Electronics Conference System　(31)
- 8.1　Construction Preparation　(31)
- 8.2　Equipment Installation　(31)
- 8.3　Quality Control　(35)
- 8.4　System Debugging　(36)
- 8.5　Self-examination and Test　(38)
- 8.6　Quality Record　(39)

9　Broadcasting System　(40)
- 9.1　Construction Preparation　(40)
- 9.2　Equipment Installation　(40)
- 9.3　Quality Control　(41)
- 9.4　System Debugging　(41)
- 9.5　Self-examination and Test　(42)
- 9.6　Quality Record　(43)

10　Information Facilities System　(44)
- 10.1　General Requirements　(44)
- 10.2　Equipment Installation　(44)
- 10.3　Quality Control　(47)
- 10.4　System Debugging　(48)
- 10.5　Self-examination and Test　(51)
- 10.6　Quality Record　(54)

11　Information Application System　(55)
- 11.1　General Requirements　(55)
- 11.2　Construction Preparation　(55)
- 11.3　Installation of the Hardware and Software　(55)
- 11.4　Quality Control　(56)
- 11.5　System Debugging　(56)
- 11.6　Self-examination and Test　(57)
- 11.7　Quality Record　(57)

12　Building Automation System　(58)
- 12.1　Construction Preparation　(58)
- 12.2　Equipment Installation　(58)
- 12.3　Quality Control　(61)
- 12.4　System Debugging　(62)
- 12.5　Self-examination and Test　(65)
- 12.6　Quality Record　(68)

13　Fire Alarm and Control System　(69)
- 13.1　Construction Preparation　(69)
- 13.2　Equipment Installation　(69)
- 13.3　Quality Control　(70)

13.4	System Debugging	(70)
13.5	Self-examination and Test	(70)
13.6	Quality Record	(71)
14	Security System	(72)
14.1	Construction Preparation	(72)
14.2	Equipment Installation	(72)
14.3	Quality Control	(74)
14.4	System Debugging	(75)
14.5	Self-examination and Test	(76)
14.6	Quality Record	(77)
15	Intelligentized Integration System	(78)
15.1	Construction Preparation	(78)
15.2	Installation of Hardware and Software	(80)
15.3	Quality Control	(80)
15.4	System Debugging	(81)
15.5	Self-examination and Test	(81)
15.6	Quality Record	(82)
16	Lightning Protection and Grounding	(83)
16.1	Equipment Installation	(83)
16.2	Quality Control	(85)
16.3	System Debugging	(86)
16.4	Self-examination and Test	(86)
16.5	Quality Record	(86)
17	Machine Room Engineering	(87)
17.1	Construction Preparation	(87)
17.2	Equipment Installation	(87)
17.3	Quality Control	(88)
17.4	System Debugging	(89)
17.5	Self-examination and Test	(89)
17.6	Quality Record	(90)
Appendix A	Records of Project Implementation and Quality Control	(91)
Appendix B	Test Records	(97)
Explanations of Wording in This Code		(128)
List of Quoted Standards		(129)

1 General Provisions

1.0.1 This code is formulated with a view to strengthening the management of the installation of intelligent building systems, guaranteeing the installation quality of intelligent building systems, and achieving advanced technology, reliable process, economy and rationality and efficient management.

1.0.2 This code is applicable to the installation of intelligent building systems in construction, renovation and extension engineering.

1.0.3 This code shall be used in association with the current national standards GB/T 50314 *Standard for Design of Intelligent Building*, GB 50300 *Unified Standard for Constructional Quality Acceptance of Building Engineering* GB/T 50339 *Code for Acceptance of Quality of Intelligent Building Systems* GB/T 50326 *The code of Construction Project Management*, GB/T 50375 *Evaluating standard for excellent quality of building engineering*, GB 50303 *Code of Acceptance of Construction Quality of Electrical Installation in Building* JGJ 46 and *Technical Code for Safety of Temporary Electrification on Construction Site*.

1.0.4 The installation of intelligent building systems shall not only comply with the requirements stipulated in this code, but also comply with those in the current relevant ones of China.

2　Terms

2.0.1　Deepening design

The process to detail the construction scheme on the basis of scheme design and technical design and also to plot the construction drawing.

2.0.2　Comprehensive pipeline

The basic platform of the building intelligent system, which is the basic channel of the construction and normal function exertion of its subsystems and also is the public pipeline for the subsystems of building intelligent system to provide the demand.

2.0.3　Hybrid fiber coaxial

The access network which takes optical fiber as the main line and coaxial cable as the distribution network.

2.0.4　Public address system

An organic entirety formed by all the broadcasting equipment and facilities serving for public places and the acoustical environment of the public coverage areas.

2.0.5　Net control unit

The device that is used for the communication between the server/workstation and the field control unit to complete the function transformation between the field control network and the IP network.

2.0.6　Building automation system

The automatic control system that utilizes automatic control technology, communication technology, computer network technology, database and image processing technology to execute comprehensive automatic monitoring, control and management over the operation, safety condition, energy use state, energy conservation and the like of various electromechanical equipment (including heating, ventilation and air conditioning, cold/heat source, water supply and drainage, power transformation and distribution, lighting and elevator/lift, etc.) affiliated to the buildings (or building complexes).

2.0.7　Intelligented integration system

The system that is formed by integrating the building intelligent systems with different functions through a unified information platform to have such comprehensive functions as information integration, resource sharing and management optimization.

2.0.8　Security system

The electronic system that integrates the subsystems such as intrusion alarm system, video security guard monitoring system and access control system and realizes the effective linkage, management and/or monitoring over the subsystems.

2.0.9　Conference system

A system that is set for completing an integral conference and is composed of the equipment or devices having all or parts of such functions as discussing, voting, identification, listening, recording and audio and video playing.

2.0.10 Self-examination and test

The activities of the constructor conducting measurement, inspection and test over the inspection items and comparing the results with the requirements of standards and specifications to determine whether all the inspection items are qualified.

3 Basic Requirements

3.1 General Requirements

3.1.1 Before installation of intelligent building systems, deepening design shall be carried out on the basis of schematic design and technical design, and the construction drawing shall be plotted.

3.1.2 The installation of intelligent building systems must be undertook by the construction organization having the corresponding qualification grade and safety production license.

3.2 Construction Management

3.2.1 The construction site management shall meet the following requirements:

  1 Coordination shall be carried out between subsystems of building intelligent system and between the building intelligent system speciality and the specialties of building engineering, and the construction progress and quality shall be guaranteed;

  2 The implementation of intelligent building systems shall accept supervision of the supervision engineer over the whole process;

  3 Without the confirmation of supervision engineer, the operation of concealed work shall not be implemented. The process examination record of concealed work shall be signed and confirmed by supervision engineer, and the concealed work acceptance table shall be filled in.

3.2.2 Construction technical management shall meet the following requirements:

  1 Under the presiding of technical director, the project department shall establish construction technical clarificaiton system that is adaptable to this project;

  2 The technical clarification materials and records shall be collected, sorted and preserved by documenter;

  3 In case of design change, it shall be consulted jointly by employer organization, design organization, supervision engineer and construction organization and shall not be implemented until the Design Change List is filled in, and it shall be checked and confirmed according to the requirements.

3.2.3 The construction quality management shall meet the following requirements:

  1 Shall determine the quality objective;

  2 Shall establish quality assurance system and quality control procedure.

3.2.4 The construction safety management shall meet the following requirements:

  1 Shall establish safety management organization;

  2 Shall meet the requirements for safety production of the state and relevant industries;

  3 Shall establish safety production system and formulate safety operation specification;

  4 Shall conduct safety production technical clarification to the working team and group prior to operation.

3.3 Construction Preparation

3.3.1 Technical preparation shall meet the following requirements:

1 Prior to construction, the deepening design shall be carried out and the plotting work of construction drawing shall be finished;

2 Construction drawing shall be reviewed and signed jointly by the employer organization, design organization and construction organization;

3 The installation of intelligent building systems shall be implemented according to the examined and approved design documents, like construction drawing;

4 The construction organization shall compile the construction organization design and special construction scheme and submit them to supervision engineer for approval;

5 Safety education and technical clarification work (including being familiar with the construction drawing, construction scheme and relevant materials, etc.) shall be conducted for construction personnel.

3.3.2 In addition to those specified in Section 3.2, Article 3.3.4 and Article 3.3.5 of the current national standard GB 50339 - 2003 *Code for Acceptance of Quality of Intelligent Building Systems*, the preparation of materials and equipment shall also meet the following requirements:

1 The materials and equipment shall be accompanied with product qualification certificate and quality inspection report, the equipment shall have product qualification certificate, quality inspection report and instruction book, etc.; the imported products shall be provided with the certificate of origin and commodity inspection certificate, certificate of quality, test report and installation, application and maintenance manual (in Chinese);

2 The brand, origin, model, specification, quantity and appearance of wire cables and equipment shall be checked, the main technical parameters, performances etc. shall meet the design requirements, the appearance shall be free from damage, the site-approaching inspection record shall be filled in, the sample of wire cable and devices shall be sealed up for safekeeping;

3 The active equipment shall be inspected by power on to confirm the equipment being normal.

3.3.3 The preparation of machines and tools, instruments and manpower shall meet the following requirements:

1 The erecting tools shall be complete and in good condition, and the electric tools shall be carried with insulation inspection;

2 The measuring instruments and tools used during the process of construction shall be calibrated according to the relevant laws and regulations of China;

3 The construction personnel shall take the post with a certificate.

3.3.4 The construction environment shall meet the following requirements:

1 The procedure handover and interface confirmation of the intelligent building systems with such specialties as building structure, building decoration, building water supply and drainage, heating, ventilation and air conditioning, building electricity and lift shall be well done;

2 The construction site shall have such conditions as water and electricity comsuption meeting the demand of normal construction;

3 The electricity for construction shall have safety protection devices, grounded reliably, and shall meet the grounding standard of safety electricity utilization;

4 The construction of the building lightning protection and grounding has been completed basically.

3.3.5 The construction preparation for all the systems included in this code shall meet those specified in Section 3.3 of this code.

3.4 Project Implementation

3.4.1 Inspection and appearance quality acceptance shall be conducted over the installation quality of engineering equipment according to such engineering design documents as construction drawing by adopting such methods as on-site observation and random inspection and test. The inspection lots shall be divided according to those specified in Article 4.0.5 and Article 5.0.5 of the current national standard GB 50300 - 2001 *Unified Standard for Constructional Quality Acceptance of Building Engineering*. The quality acceptance records shall be filled in according to the corresponding requirements in appendixes when inspecting and shall also be kept properly.

3.4.2 The laying route of wire slots and wire cables of the subsystems in the intelligent building system shall be consistent, the wire slots and wire cables of the subsystems should be laid synchronously, the wire cables shall be left with surplus according to requirements and the terminals of wire cables shall be done with such protective measures as sealing and moistureproofing.

3.4.3 The wire slots and wire cables shall be marked clearly.

3.5 Quality Assurance

3.5.1 In addition to the requirements of Article 3.2.1 and Article 3.2.2 in the current national standard GB 50339 - 2003 *Code for Acceptance of Quality of Intelligent Building Systems*, the approaching quality test for material, machine and tool, equipment shall also meet the following requirements:

    1 The site acceptance shall be carried out in accordance with the contract documents and engineering design documents, be with the written records and the participant signature, and be confirmed by the supervision engineer or the acceptance personnel of employer organization;

    2 The appearance, specification, model, quantity, place of origin etc. of the materials and equipment shall be inspected rechecked;

    3 The main equipment and materials shall be provided with the manufacturer's quality compliance certification document and performance test report;

    4 The quality inspection of equipment and materials shall include such items as safety, reliability and electromagnetic compatibility (EMC), and shall be issued with corresponding test report by the manufacturer.

3.5.2 In addition to those specified in Article 3.0.1 of the current national standard GB 50300 - 2001 *Unified Standard for Constructional Quality Acceptance of Building Engineering*, the installation quality assurance of the building intelligent subsystems shall also meet the following requirements:

    1 The installation and debugging personnel shall have the corresponding professional qualification or special qualification;

    2 The operation personnel shall be qualified through on-the-job training and shall hold the accreted certificate;

    3 The instruments, meters and measuring devices shall be provided with the examination and calibration certificates within validity period.

3.5.3 The installation quality test of the subsystems shall meet the following requirements:

1 The installation quality test of the subsystems shall comply with the current national or professional standards;

2 After completing the equipment installation, the construction organization shall conduct self-examination over the system. During the self-examination, the test items shall be tested item by item and the relevant records shall be well done.

3.5.4 The test of intelligent building systems shall meet the following requirements:

1 The quality of the subsystem interfaces shall be inspected according to the following requirements:

1) As for all the interfaces, the interface suppliers shall submit the interface specification and interface test outline;

2) The interface specification and interface test outline should be examined and approved with the participation of construction organization of intelligent building systems when signing the contract;

3) The construction organization shall implement inspection in accordance with the test outline and shall ensure the installation quality of system interface.

2 The construction organization shall organize the personnel concerned to formulate the system test scheme in accordance with the corresponding requirements of contract technical documents, design documents and this code.

3 The conclusion and treatment method of system test shall meet those specified in Article 3.4.4 of the current national standard GB 50339 - 2003 *Code for Acceptance of Quality of Intelligent Building Systems*.

4 The test record shall be filled in according to Appendix B of this code.

3.5.5 The quality inspection on software products shall meet the following requirements:

1 The application licence and application scope shall be checked;

2 The user application software, designed software configuration, interface software etc. shall be carried out with functional test and system test and be provided with complete documents (including program structure instruction, installation and debugging instruction and operation and maintenance instructions, etc.).

3.6 Finished Product Protection

3.6.1 Aiming at the characteristics of equipment in different subsystems, finished product protection measures shall be formulated.

3.6.2 The equipment having been installed on site shall be taken with such necessary protective measures as packing, covering and separating and also shall be kept avoiding impact and damage.

3.6.3 The equipment stored at the construction site shall be taken with such protective measures against dust, moisture, impact, smash, pressure and theft.

3.6.4 During construction, the main switch of the equipment power supply shall be turned off on meeting lightning, overcast and rainy and wet weather or a long time of out of service.

3.6.5 The protection for the software and system configuration shall meet the following requirements:

1 The alteration of software and system configuration shall be well done with records;

2 During the process of debugging, the software shall be backed up every day, the backup contents shall include system softwares, database, configuration parameters and system images;

3 The backup files shall be saved in independent storage devices;

4 The login password of the system equipment shall be managed by specific personnel and shall not be disclosed;

5 The computer shall be locked in case of no operator.

3.7 Quality Record

3.7.1 The inspection records of construction site quality management shall be filled in accordance with Table A.0.1 of the current national standard GB 50339 - 2003 *Code for Acceptance of Quality of Intelligent Building Systems*.

3.7.2 The site-approaching inspection records of equipment and materials shall be filled in Table B.0.1 of the current national standard GB 50339 - 2003 *Code for Acceptance of Quality of Intelligent Building Systems*.

3.7.3 The concealed work inspection records shall be filled in Table B.0.2 of the current national standard GB 50339 - 2003 *Code for Acceptance of Quality of Intelligent Building Systems*.

3.7.4 In case of the alteration of examination and verification records, Table B.0.3 of the current national standard GB 50339 - 2003 *Code for Acceptance of Quality of Intelligent Building Systems* shall be filled in.

3.7.5 The installation quality and appearance quality acceptance records of the engineering shall be filled in Table B.0.4 of the current national standard GB 50339 - 2003 *Code for Acceptance of Quality of Intelligent Building Systems*.

3.7.6 The equipment unpacking inspection records shall be filled in Table A.0.1 of this code.

3.7.7 The design change records shall be filled in Table A.0.2 of this code.

3.7.8 The engineering negotiation records shall be filled in Table A.0.3 of this code.

3.7.9 The joint review records of drawings shall be filled in Table A.0.4 of this code.

3.7.10 The subitem project quality test records of intelligent building systems shall be filled in Table C.0.1 of the current national standard GB 50339 - 2003 *Code for Acceptance of Quality of Intelligent Building Systems*.

3.7.11 The subsystem test records shall be filled in Table C.0.2 of the current national standard GB 50339 - 2003 *Code for Acceptance of Quality of Intelligent Building Systems*.

3.7.12 The compulsory measure provisions test records shall be filled in Table C.0.4 of the current national standard GB 50339 - 2003 *Code for Acceptance of Quality of Intelligent Building Systems*.

3.7.13 The system (subsection) project test records shall be filled in Table C.0.4 of the current national standard GB 50339 - 2003 *Code for Acceptance of Quality of Intelligent Building Systems*.

3.7.14 The pretest records shall be filled in Table B.0.1 of this code.

3.7.15 The inspection lot test records shall be filled in Table B.0.2 of this code.

3.7.16 The system debugging records shall be filled in Table B.0.3 of this code.

3.7.17 The quality records of all the subsystems included in this code shall meet those specified in Section 3.7 of this code.

3.8 Safety, Environmental Protection and Energy Conservation Measures

3.8.1 The safety measures shall meet the following requirements:

1 Safety clarification shall be carried out before and during construction;

2 The electricity utilization on construction site shall comply with the relevant requirements of the current professional standard JGJ 46 *Technical Code for Safety of Temporary Electrification on Construction Site*;

3 When measuring the optical cable by adopting optical power meter, direct observation by naked eyes is disallowed;

4 As for aerial operation, the scaffolds and ladders shall be safe and reliable, the ladders shall be taken with anti-skidding measures and the operation by two persons on one ladder is disallowed;

5 On meeting strong wind or heavy thunderstorm weather, the outdoor high-altitude installation operation shall not be performed;

6 The one who enters into the construction site shall wear safety helmet; in case of aerial operation, the operation personal shall fasten the safety belt;

7 At the construction site, attention shall be drawn to fire protection and effective fire-fighting equipment shall be equipped;

8 Before installing or cleaning the active equipment, this equipment shall be powered off in advance. The live equipment shall not be cleaned or wiped by using liquid or the wet cloth;

9 The equipment shall be placed stably, and water or moisture shall be prevented from entering into the active hardware equipment;

10 It shall be confirmed that the power supply voltage is consistent with the rated voltage of electric equipment;

11 When the hardware equipment is working, its enclosure shall not be opened;

12 When replacing the wiring board, static protective gloves should be used;

13 The power line shall be avoided being tramped, pulled or dragged.

3.8.2 The environmental protection measures shall not only comply with the relevant requirements of the current professional standard JGJ 146 *Standard of Environment and Sanitation of Construction Site*, but also meet the following requirements:

1 Rubbishes and waste materials at the site shall be stacked at designated place, timely cleared or reclaimed, and shall not be thrown at will;

2 As for the noise from the site construction machines and tools, corresponding measures shall be taken to reduce it to the utmost;

3 Measures shall be taken to control the dust pollution during the construction process.

3.8.3 The energy conservation measures shall meet the following requirements:

1 The materials shall be saved and consumption shall be reduced, and the macroscopic energy conservation awareness shall be strengthened;

2 The energy-saving type lighting fixtures shall be selected to reduce the power consumption of lighting and improve lighting quality;

3 The electric tools for construction shall be timely repaired, overhauled, maintained and replaced, the system failures shall be timely eliminated and the energy consumption shall be reduced.

4 Comprehensive Pipeline

4.1 General Requirements

4.1.1 The power wire cable and signal wire cable must not be laid in one wire pipe.

4.1.2 The construction of the wire cables of generic cabling system shall meet those specified in Chapter 5 of this code.

4.2 Construction Preparation

4.2.1 Before construction, the bridges and wire pipes of all systems shall be carried out with integrated layout and arrangement, and the construction drawing of intelligent system shall be plotted after deepening design and shall be approved through joint review.

4.2.2 The construction organization shall cooperate with the engineering general contractor and design organization to complete the comprehensive pipeline layout and arrangement design of various professions.

4.2.3 The material preparation shall meet the following requirements:

    1 The specification and type of bridge, wire pipe and wire cable shall meet the design requirements and shall be provided with the product qualification certificates and test reports.

    2 Parts of bridge and wire pipe shall be complete, with smooth surface and complete coating and be rustless.

    3 The metal conduit shall be free from such defects crack, burr, fin, pin hole and bubble, be of uniform wall thickness and even conduit orifice; the insulating conduit and fittings shall be in good condition and be with antiflaming marks on surface.

    4 The wire cables shall be carried out with the inspection on connecting, disconnecting electricity and insulation between cables.

4.3 Pipeline Installation

4.3.1 Bridge installation shall meet the following requirements:

    1 Anti-corrosive measures shall be taken at the cutting and drilling sections of bridge;

    2 Bridge shall be even without distortion and deformation, and its inner wall shall be burr free, all the accessories shall be installed ready, the nut of fastener shall be set at the outer side of bridge, the bridge interface shall be straight, even and tight, the cover plates shall be complete, flat and smooth;

    3 The bridge shall be set with compensation devices at the place passing through the deformation joints (including settlement joint, expansion joint and seismic joint, etc.) of building, the protective ground wire and the wire cables in bridge shall be reserved with compensating surplus;

    4 The joint of bridge with box, case, cabinet and the like shall adopt the feet fitted connection or flanged edge connection and shall be fixed by screw, the terminals shall be blocked;

5 The distance from the bottom of horizontal bridge to the ground should not be less than 2.2 m, the distance from its top to the floor slab should not be less than 0.3 m, the distance from bridge to beam should not be less than 0.05 m, and the spacing between bridge and power cable should not be less than 0.5 m;

6 Where the bridge is parallel to or cross with various pipelines, their minimum clearance shall meet those specified in Table 12.2.1-2 in Article 12.2.1 of the current national standard GB 50303-2002 *Code of Acceptance of Construction Quality of Electrical Installation in Building*;

7 Bridges and pipeline holes laid in vertical shaft or passing through different fire compartments shall be provided blocked with firestops;

8 Fittings, like elbows and tees, should be adopted the finished products manufactured by bridge manufacturer and should not be processed or fabricated on site.

4.3.2 Installation of supports and hangers shall meet the following requirements:

1 The spacing of the supports and hangers installed in the straight section should be 1.5m~2.0m, and the spacing of supports and hangers in one straight section shall be uniform;

2 Supports and hangers shall be installed within a scope of no larger than 0.5 m at the port, branch or turn of bridge;

3 Supports and hangers shall be straight and even without obvious distortion, their welding shall be firm without obvious deformation, the weld seam shall be uniform, flat and smooth, and cut place shall be free from flanged edge and burr;

4 Supports and hangers shall be securely connected and fixed by adopting expansion bolts and shall also be assembled with spring washers;

5 Supports and hangers shall be conducted with anticorrosive treatment;

6 Where round steel is adopted as the hanger, the anti-sway supports shall be installed at the turns of bridge and also every other 30 m in the straight section of bridge.

4.3.3 Wire pipe installation shall meet the following requirements:

1 Conduit laying shall be keep clean and dry inside conduit, the conduit orifice shall be taken with protective measures and be carried out with plugging treatment;

2 The exposed wire pipes shall be straight and even and also be lined neatly;

3 The exposed wire pipes shall be fixed by pipe clips, the pipe clips shall be installed firmly and their arrangement shall meet the following requirements:

    1) Pipe clips shall be installed within the scope of 150 mm~500 mm at the terminal and the midpoint of elbow;

    2) Pipe clips shall be installed within the scope of 150 mm~500 mm away from the edges of box, case and cabinet, etc.;

    3) Pipe clips shall be installed uniformly in the middle straight section. The maximum distance between pipe clips shall meet those specified in Table 14.2.6 of the current national standard GB 50303-2002 *Code of Acceptance of Construction Quality of Electrical Installation in Building*;

4 The bending radius at the turn of wire pipe shall not be less than the minimum allowable bending radius of the wire cable that is put into the wire pipe and shall not be less than 6 times of the outside diameter of this pipe; where the outside diameter of concealed pipe is larger than 50mm, the bending radius shall not be less than 10 times of the outside diameter of this pipe;

5 The buried depth of the concealed wire pipe in masonry shall not be less than 15 mm and that of concealed wire pipe in cast-in-place concrete floor slab shall not be less than 25 mm, the spacing between those parallelly laid wire pipes shall not be less than 25 mm;

6 Where the wire pipe is connected to control case, wiring case, junction box or the like, locknut shall be adopted to firmly fix the pipe orifice;

7 The wire pipe shall be installed with protective sleeve when passing through wall or floor slab, the sleeve through wall shall be level with the wall surface, the upper opening of sleeve through floor should be 10 mm ~ 30 mm higher than the floor and its lower opening shall be level with the floor;

8 Where the wire pipe connected with equipment is led out of ground, the distance from pipe mouth to ground should not be less than 200 mm; where the wire pipe is led into floor type case or cabinet underground , it should be 50 mm higher than the inner bottom surface of the case or cabinet;

9 Wire pipe shall be set with marks at both ends, it shall be free from obstacle inside and shall also be threaded with strip line;

10 The wire pipe in suspended ceiling should be fixed by using separate support and hanger, and the support and hanger shall not be erected on the keel or other pipelines;

11 Where the wire pipe passes through the deformation joint of building, it shall be set with compensation device;

12 The galvanized steel pipe should adopt threaded connection, the joint of galvanized steel pipe shall be adopted with dedicated grounding wire clip to fix the jumper, and the cross section of jumper shall not be less than 4 mm^2;

13 Non-galvanized steel pipe shall be adopted sleeve pipe welding, and the sleeve pipe length shall be 1.5 ~ 3.0 times of the pipe diameter;

14 Welded steel pipe shall not be bent at weld, and the bending place shall be free from such phenomena as bending and folding; the galvanized steel pipe shall not be bent by heating;

15 The connection of muff-coupling clamping steel pipe shall meet the following requirements:

1) The plating on the outer wall of steel pipe shall be in good condition, the pipe mouth shall be even and smooth without deformation;

2) Sealing measures shall be taken at the joint of muff-coupling clamping steel pipe;

3) Where the diameter of the muff-coupling clamping steel pipe is larger than or equal to 32 mm, no less than 2 set tight and fixed screws shall be set at each end of the connecting sleeve.

16 The laying of outdoor wire pipe shall meet the following requirements:

1) As for the outdoor buried wire pipes, the buried depth should not be less than 0.7m and the wall thickness shall be larger than or equal to 2 mm; if the wire pipe is buried under the hard road surface, it shall be added with steel sleeve, and the manhole and handhole wells shall be with drainage measures;

2) The wire pipe in and out of building shall be done with waterproofing gradient, and the gradient should not be larger than 15‰;

3) One wire pipe section should not be made with S curve within a short distance;

4) Where the wire pipe enters into underground building, it shall be adopted with

waterproofing sleeve and shall be done with sealing and waterproofing treatment.

4.3.4 Installation of wire boxes shall meet the following requirements:

1 Where the steel conduit enters into box (case), one conduit shall pass through a hole, and the connection between conduit and box (case) shall be adopted the claw-type screwed joint pipe connection, and the conduit shall be tightly locked, the inner wall of conduit shall be smooth and clean for the convenience of guiding wire.

2 If the wire pipeline has any one of the following conditions, pull box or junction box shall be added at middle and its position shall be convenient for guiding wire:

1) The pipeline has no bend in per length exceeding 30m;

2) The pipeline has only one bend in per length exceeding 20m;

3) The pipeline has only two bends in per length exceeding 15m;

4) The pipeline has only three bends in per length exceeding 8m;

5) Where the wire cable pipeline is laid vertically, the cross section of the insulated wire cable in pipe should be less than 150mm^2; where the pipeline length is larger than 30m, the pull boxes for purpose of fixing shall be added;

6) Embedded box at information point should not be used as line passing box at the same time.

4.4 Wire Cable Laying

4.4.1 The wire cable shall be provided with permanent labels for waterproofing and friction resistance at both ends, and the characters on the labels shall be distinct and accurate.

4.4.2 Wire cables in a pipe shall not be twisted or buckled and shall be without joints.

4.4.3 The minimum allowable bending radius of wire cable shall meet those specified in Table 12.2.1-1 of the current national standard GB 50303 - 2002 *Code of Acceptance of Construction Quality of Electrical Installation in Building*.

4.4.4 Connection between outlet of wire pipe and connecting terminal of equipment shall be adopted metal hose, the length of metal hose should not exceed 2 m, and the wire shall not be exposed.

4.4.5 Wire cables in bridge shall be arranged neatly and not be twisted or buckled; the wire cables shall be fixed by banding at the positions in and out of the bridge and at their turns; the spacing between the banding and fixing points of wire cable in vertical bridge should not be larger than 1.5 m.

4.4.6 Where the wire cable passes through the deformation joints of building, it shall be retained with appropriate compensating surplus.

4.4.7 In addition to complying with those specified in this code, the laying of wire cables the relevant requirements of the current national standards GB 50200 *Technical code for Regulation of CATV system*, GB 50303 *Code of Acceptance of Construction Quality of Electrical Installation in Building* and GB 50348 *Technical Code for Engineering of Security and Protection System*.

4.5 Quality Control

4.5.1 The dominant items shall meet the following requirements:

1 The bridges laid in vertical shaft or passing through different fire compartments as well as the holes of wire pipes shall be provided with blocked for fire protection;

2 The place where the bridge or wire pipe passes through the deformation joints of building shall be equipped with compensation device, and the wire cable shall be retained with surplus;

3 The wire cable shall be provided with permanent labels for waterproofing and friction resistance at both ends, and the characters on the labels shall be distinct and accurate;

4 The bridge, wire pipe and junction box shall be grounded reliably; where integrated grounding is adopted, the ground resistance shall not be larger than 1 Ω.

4.5.2 General items shall meet the following requirements:

1 The bridge shall be taken with anticorrosive measures after being cut and drilled with holes, and the supports and hangers shall be carried out with anticorrosive treatment;

2 The wire pipe shall be equipped with marks at both ends and shall be threaded with strip line;

3 Where the wire pipe is connected to control case, wiring case, pull box or the like, locknut shall be adopted, and the wire pipe, case and box shall be firmly fixed;

4 The pipe in suspended ceiling should be fixed by using separate support and hanger, and the support and hanger shall not be erected on the keel or other pipelines;

5 Sealing measures shall be taken at the joint of steel pipe with tapered adapter;

6 Bridge shall be firmly installed, straight and even, without distortion;

7 Wire cables in bridge or wire pipe shall not be twisted or buckled and shall be without joints.

4.6 Self-examination and Test

4.6.1 The bridge and wire pipe shall be inspected for their specification, position, bending and flattening degree, bending radius, connection, bridge ground wire, corrosion protection, pipe box fixation, pipe orifice treatment, protective layer and welding quality, etc. The bent pipe and connection accessories shall be of uniform radian and shall be free from such defects as folding, depression, crack, bending or flattening and dead turn, and the welds of pipe shall be at the outer side.

4.6.2 According to the requirements of deepening design documents, the specification, type, labeling and laying quality of wire cable shall be inspected.

4.6.3 After the construction of concealed work is completed, the concealed work record sheet shall be filled in.

4.6.4 After the concealed work is qualified in acceptance, Table B.0.2 of this code shall be filled in.

4.6.5 As for the ground resistance test of bridge and wire pipe, Table B.0.27 of this code shall be filled in.

5 General Cabling System

5.1 Construction Preparation

5.1.1 The preparation of materials and equipment used in engineering shall meet the following requirements:

1 As for the approaching test of wire cables, the electrical property indicators of cable shall be randomly inspected, and the relevant records shall be noted down;

2 As for the approaching test of optical fibers, the performance indicators of the optical fibers of optical cables shall be randomly inspected, and the relevant records shall be noted down.

5.2 Wire Cable Laying and Equipment Installation

5.2.1 In addition to complying with those specified in Section 4.4 of this code, the laying of wire cables shall also meet the following requirements:

1 The wire cable shall be laid naturally, straightly and evenly and shall not suffer extrusion and any damage from external force;

2 The wire cable shall be laid with a surplus no less than 0.15 mm;

3 The length of 4-pair twisted cable led from the distribution frame to each information port of working area shall not be larger than 90 m;

4 The tension and other protective measures for the laying of wire cables shall meet the construction requirements of product manufacturer;

5 The bending radius of wire cable should meet the following requirements:

    1) The bending radius of the unshielded 4-pair twisted cable should not be less than 4 times of the outside diameter of cable;

    2) the The bending radius of the shielded 4-pair twisted cable should not be less than 8 times of the outside diameter of cable;

    3) The bending radius of the main paired cable should not be less than 10 times of the outside diameter of cable;

    4) The bending radius of optical cable should not be less than 10 times of the outside diameter of optical cable;

6 The clear distance between wire cables shall be in accordance with those specified in Article 5.1.1 of the current national standard GB 50312 - 2007 *Code for Engineering Acceptance of Generic Cabling System*;

7 Where the indoor optical cables are laid in bridge, cushion cover should be added at the banding and fixing place;

8 During the laying construction of wire cables, stable temporary wire number labels shall be installed on site; prior to laying wire cables on distribution frame or driving the module, permanent wire number labels shall be installed;

9 When wire cable pass the bridge or pipeline turning, it shall be guaranteed that the wire cable is closely fitted to the bottom, not suspended and is free from traction force. At the turn of

bridge, the wire cable shall be fixed by banding or by other modes;

    10 When threading in the line passing box that is nearest to the information point, a surplus no less than 0.15 mm should be reserved.

5.2.2 The installation elevation of information outlet shall meet the design requirements, the horizontal distance installed between this outlet and the power socket shall be in accordance with those specified in Article 5.1.1 of the national standard GB 50312-2007 *Code for Engineering Acceptance of Generic Cabling System*. Where there is no requirement in the design, installation elevation of information outlet shall be equal to that of the power socket.

5.2.3 Wire cables in cabinet shall be banded on the cable management arms on both sides of the cabinet and shall be arranged neatly and beautifully, the distribution frame shall be installed firmly, and the information point labeling shall be accurate.

5.2.4 Fiber distribution frame (board) should be installed at the top of cabinet, and the switchboard should be installed between the copper cable distribution frame and fiber distribution frame (board).

5.2.5 Local equipotential terminal board shall be set in the distribute room and the cabinet shall be grounded reliably.

5.2.6 The jumper wire shall be connected with the relevant equipment through the cable management arm, and the wire cables shall be arranged neatly inside and outside the cable management arm.

5.3 Quality Control

5.3.1 Quality control shall comply with the relevant requirements of the current national standards "Code for Engineering Acceptance of Generic Cabling System" GB 50312 and "Code for Acceptance of Quality of Intelligent Building Systems" GB 50339.

5.4 Channel Test

5.4.1 The technical index of the permanent link of wire cable shall meet the relevant requirements of the current national standard GB 50311 *Code for Engineering Design of Generic Cabling System*.

5.4.2 The electrical performance test of cable and the performance test of optical fiber system shall meet the relevant requirements of the current national standard GB 50312 *Code for Engineering Acceptance of Generic Cabling System*.

5.5 Self-examination and Test

5.5.1 The inspection items and contents in relation to the laying of wire cable and the installation of distribution wire equipment shall be in accordance with those specified in Table 5.5.1.

Table 5.5.1 **Inspection Items and Contents on Wire Cable Laying and Distribution Wire Equipment Installation**

Phase	Inspection item	Inspection content	Inspection mode
Equipment installation	Distribute room and equipment cabinet	1. Specification and appearance; 2. Installation perpendicularity and levelness; 3. The paint shall not shed off and the marking shall be intact;	Inspection following construction

Table 5.5.1(continued)

Phase	Inspection item	Inspection content	Inspection mode
Equipment installation	Distribute room and equipment cabinet	4. Various screws must be fastened; 5. Seismic strengthening measures; 6. Grounding measures; 7. Power supply measures; 8. Heat dissipation measures; 9. Lighting measures	Inspection following construction
	Distribution wire equipment	1. Specification, position and quality; 2. Various screws must be tightened up; 3. The identification is complete; 4. The installation shall meet the process requirements; 5. The shielding layer shall be reliably connected	Inspection following construction
Wire cable laying (in building)	Concealed laying of wire cable (including such mode as concealed pipe, wire slot and floor)	1. Specification, route and position of wire cable; 2. Shall meet the process requirements for wire cable laying; 3. Installation of pipe chase shall meet the process requirements; 4. Grounding measures	Concealed work visa
Wire cable laying (between buildings)	Wire cable in pipeline	1. The hole position and hole diameter of the used pipe hole; 2. Specification of wire cable; 3. Installation position and route of wire cable; 4. Protection facilities of wire cables	Concealed work visa
	Wire cable in tunnel	1. Specification of wire cable; 2. Installation position and route of wire cable; 3. Installation and fixation mode of wire cable	Concealed work visa
	Miscellaneous	1. Spacing between the wire cable route and other special pipelines; 2. Equipment installation and construction quality of equipment room	Inspection following construction or concealed work visa
Cable terminating	Information outlet	Meet the process requirements	Inspection following construction
	Wiring component	Meet the process requirements	
	Optical fiber socket	Meet the process requirements	
	Various jumper wires	Meet the process requirements	

5.5.2 The test items and contents of generic cabling system shall be in accordance with those specified in Table 5.5.2.

Table 5.5.2 Test Items and Contents of the System

Inspection item	Inspection content	Inspection mode
Basic electrical performance test of cable	1 Connection diagram; 2 Length; 3 Attenuation; 4 Near-end crosstalk (both ends shall be tested); 5 Connection conditions of cable shielding layer; 6 Other technical indexes	Self-examination
Characteristic test of optical fiber	1 Attenuation; 2 Length	Self-examination

5.6 Quality Record

5.6.1 In addition to complying with those specified in Section 3.7 of this code, the quality record of the generic cabling system also shall comply with the relevant requirements of the current national standard GB 50312 *Code for Engineering Acceptance of Generic Cabling System*.

6 Information Network System

6.1 Construction Preparation

6.1.1 The construction organization shall complete the planning and configuration scheme of the information network system according to the requirements of design document and also shall be approved through the joint review of the design organization, construction organization and employer organization.

6.1.2 The special products for system safety must have the sale permit of special products for computer information system safety examined, approved and issued by the computer management and supervision department of the Ministry of Public Security.

6.1.3 The information network system machine room shall be constructed and completed integrally.

6.2 Equipment and Software Installation

6.2.1 Equipment installation of information network system shall meet the following requirements:

1 The installation position shall meet the design requirements and the installation shall be stable, firm and convenient for operation and maintenance;

2 Equipment installed in cabinet shall be with ventilation and heat dissipation measures, and the connection between the internal connectors and equipment shall be firm;

3 The equipment with load-bearing requirement larger than 600 kg/m^2 shall be manufactured with equipment base separately and shall not be directly installed on anti-static electricity floor;

4 As for the equipment with serial number, the serial number of equipment shall be registered;

5 Active equipment shall be carried out with the power-on inspection and the equipment shall work normally;

6 The jumper connection shall be normative, the wire cables shall be arranged in order and shall be set with correct and firm labels;

7 The equipment installation cabinet shall be posted with the schematic diagram for equipment system wiring.

6.2.2 Installation of software system shall meet the following requirements:

1 The equipment shall be installed with corresponding software system according to the design document, and the system installation shall be complete;

2 The legitimate software technical manual shall be provided;

3 The server shall not be installed with any software irrelevant to this system;

4 Both the operating system and the antivirus software shall be set to the auto update mode;

5 After installation, the software system shall be able to normally start up, operate and quit;

6 After the network security inspection, the server may be connected with the internet under

the protection of safety system and also shall upgrade the operating system and antivirus software and update the corresponding patch programs.

6.2.3 The safety of the installation and operating of equipment and software shall be in accordance with those specified in Article 11.3.7 of this code.

6.3 Quality Control

6.3.1 The dominant items shall meet the following requirements:

1 The inspection on the computer network system shall be in accordance with those specified in Article 5.3.3 and Article5.3.4 of the current national standard GB 50339 – 2003 *Code for Acceptance of Quality of Intelligent Building Systems*;

2 The system test and sample quantity of inspection shall meet the design requirements of information network system;

3 The system configuration shall conform to the planning and configuration scheme approved through examination and verification, and complete record shall be available.

6.3.2 General items shall meet the following requirements:

1 Functions of computer network, such as fault tolerance and network management, shall be tested in accordance with those specified in Article 5.3.5 and Article 5.3.6 of the current national standard GB 50339 – 2003 *Code for Acceptance of Quality of Intelligent Building Systems*, and the records shall be filled in earnestly;

2 Extensibility, fault tolerance and maintainability of the software system shall be inspected;

3 The network security management system, the environmental conditions of machine room, and the leakproof and confidential measures shall be inspected.

6.4 System Debugging

6.4.1 The debugging preparation shall meet the following requirements:

1 Inspection on the installation and connection work of hardware and software shall be completed and the power on work of equipment shall be normal;

2 The network planning and configuration scheme shall be completed and also be approved through joint review;

3 The formulation of the network security scheme shall be completed and also be approved through joint review;

4 The formulation of joint debugging scheme of computer network system, application software and information security system shall be completed and also be approved through joint review;

5 Before system debugging, the preparation work such as relevant data and aggressive software sample for debugging of information network system shall be well done.

6.4.2 Debugging of information network system shall meet the following requirements:

1 Network management system software shall be installed at the network management workstation and the highest management authority shall be configured;

2 Network segments and routes shall be divided according to the network planning and configuration scheme, and the network equipment shall be configured and connected;

3 The operating state, operating efficiency and operating log of the system shall be checked

every day and the error, if any, shall be modified;

4 Address of each equipment in the network shall conform to the specifications and the configuration scheme, while the address should not be directly and automatically searched and established by the network management software;

5 Each intelligent subsystem should be distributed with independent network segments;

6 Inspection shall be conducted on basis of the network planning and configuration scheme and the design requirements shall be met.

6.4.3 Debugging and testing of application software shall meet the following requirements:

1 The parameter configuration of application software and testing of software function shall be conducted in accordance with the configuration plan, function instructions and operating instructions, and the relevant records shall be made;

2 Reliability, safety, recoverability, self-examination function and other contents of the software shall be tested, and the relevant records shall be made;

3 Debugging shall be conducted according to the actual cases and actual data used by the system, and the system processing results shall be correct;

4 In the testing of application software system, the following requirements shall be met and the test results shall be recorded:

1) The functional test shall be carried out, including whether the installation is successful, and the each application functions shall be tested one by one by using actual cases;

2) Performance test, including response time, throughput, internal memory and secondary memory area, and processing accuracy of all application functions, shall be carried out;

3) The document test, including the clearness and accuracy of user's documents, shall be carried out;

4) Reliability test shall be carried out;

5) Interconnectivity test shall be carried out, and the interconnectivity between multiple systems shall be inspected;

6) Consistency test shall be carried out after software modification, and the modified software shall meet the design requirements of system.

5 The test of operation interface, data capacity, extensibility and maintainability shall be carried out for the application software as required, and the test process and results shall be recorded.

6.4.4 Debugging and testing of network security system shall meet the following requirements:

1 Software configuration of network security system shall be inspected and shall meet the design requirements;

2 Attack test shall be carried out on basis of the network security scheme, and the relevant records shall be made;

3 The site, power distribution, grounding, wiring, electromagnetic leakage, access control management and the like shall be inspected and shall meet the those specified in the system design;

4 The security debugging and test of network layer shall meet the following requirements:

1) The firewall shall be carried out with simulated attack test;

2) The management and control of the internet access shall be carried out by using acting server;

3) The test shall be carried out according to the configured network segments that are interconnected and isolated as required in the design;
　　4) Anti-virus system shall be used to carry out permanent test and the viral transmission shall be simulated based on the network security scheme, and it will be regarded as qualified if the test is correct and the anti-virus operation is performed;
　　5) Where intrusion testing system is used, the simulated attack shall be carried out on basis of the network security scheme; and it will be regarded as qualified if the intrusion testing system is able to discover and perform blocking;
　　6) Where content filtering system is used, it shall be able to block the access to limited website or content but allow normal proceeding of the non-limited access to website or content.
　5　The security debugging and test of system layer shall meet the following requirements:
　　1) The configuration of the operating system and file system shall meet the design requirements;
　　2) The system management provisions shall be formulated and enforced strictly, and the management provisions shall also be improved in appropriate time;
　　3) Configuration of the server shall conform to those specified in 6.2.2 of this code;
　　4) Auditing system shall be used to record the intrusion attempts and the recording conditions of audit log shall be checked in appropriate time to make immediate treatment.
　6　The security debugging and test of application layer shall meet the following requirements:
　　1) Identity authentication, password transmission management provisions and technicals meeting the requirement of network security scheme shall be formulated;
　　2) On the basis of identity authentication, the resource authorization list shall be formulated and also be improved in appropriate; it shall achieve that the users can correctly access the authorized resource and cannot access the unauthorized resource;
　　3) The integrity and confidentiality of data during storage, application and transmission shall be checked and improvement shall be made according to the testing conditions;
　　4) Access to the application system shall be recorded.

6.4.5　During the debugging process of information network system, corresponding records shall be filled in timely and the following requirements shall be met:

　1　When conducting every time reconfiguration or parameter modification, the change plan shall be filled in; after reconfiguration or parameter modification, corresponding records shall be updated;

　2　After the equipment and software parameters are well configured and the equipment and software operate normally, inspection, correction and perfection shall be carried out according to the function plan and design table to reach the design requirements.

6.4.6　After the parameters of network equipment, server and software system are well configured, the connection condition and safety test of the system shall be inspected and shall meet the following requirements:

　1　The softwares such as operation system, antivirus software and firewall software shall be set into the operation mode of automatic download, installation and updating;

　2　The network route, network segment division and network address shall be clearly filled in and appropriate rights shall be configured for test users;

3 The configuration, realization function and operation condition of application software system shall be clearly filled in and appropriate rights shall be configured for the test users.

6.4.7 Debugging and test for security of information network system shall meet the following requirements:

1 During construction, the system software shall be backed up every day and the backup files shall be stored in independent storage devices;

2 The personnel other than those allocated for this system shall not alter the installation and configuration of this system.

6.5 Self-examination and Test

6.5.1 Inspection on information network system shall meet those specified in Section 6.3 of this code.

6.5.2 Inspection on system documents shall meet the following requirements:

1 Documents for the configuration scheme of network system, parameter configuration of network elements and connection inspection records shall be complete;

2 Documents for configuration scheme, configuration description and inspection record of application software shall be complete;

3 Documents for configuration scheme, attack test record and inspection record of security system shall be complete.

6.5.3 Attacking softwares used for the testing of network security system as well as their carriers shall be kept properly.

6.6 Quality Record

6.6.1 The configuration table of network equipment shall be filled in Table A.0.5 of this code.

6.6.2 Configuration table of application software system shall be filled in Table A.0.6 of this code.

6.6.3 Debugging record of network system shall be filled in Table B.0.4 of this code.

7 Satellite TV Reception and Cable TV System

7.1 Construction Preparation

7.1.1 The construction organization shall obtain the satellite TV reception and cable TV system engineering construction qualification certificate issued by the relevant functional department of the nation or of the industry or speciality.

7.1.2 Prior to construction of satellite TV reception and cable TV system engineering, information such as corresponding field investigation, design documents and drawings shall be available. The construction shall be carried out in accordance with the design drawing.

7.1.3 In addition to those specified in Article 3.3.2 of this code, the equipment and material preparation also shall meet the following requirements:

 1 All active equipment shall be carried out with power-on inspection;

 2 The main equipment and material shall be selected those products with the effective identification labeling issued by the State Administration of Broadcasting, Film and Television or the inspection organization with relevant qualification certificate.

7.1.4 Concealed pipes in buildings shall meet the technical requirements of Section 4.3 in the current professional standard GY 5078 - 2008 *Technical Regulations on Safety of CATV Distribution Network Projects*.

7.2 Equipment Installation

7.2.1 Installation of satellite receiving antenna shall meet the following requirements:

 1 As for the installation of satellite antenna pedestal, the fabrication of the base shall be carried out at the same time on the poured concreting layers of civil work according to the position and dimension given in the design drawing, the foundation bolts in the base shall be connected with the top steel bars of the storied building by welding and also shall be connected with the ground grid, and the grounding resistance of antenna base shall be less than 4Ω;

 2 In front of the reception of antenna, there shall be no shelter;

 3 The required reception frequency shall be free from microwave interference;

 4 After determining the optimal orientation, the receiving antenna shall be firmly installed;

 5 The antenna adjusting mechanism shall be flexible and continuous, the locking device shall be convenient and firm and shall be provided with corrosion-resistant measures and dust-proof sheath;

 6 The satellite receiving antenna shall be within the protection scope of lightning arrester, the lightning arrester shall be possessed of good grounding system and the grounding resistance shall be less than 4Ω;

 7 The grounding of lightning arrester shall be realized by an independent wire, and the lightning protection grounding shall not share the indoor grounding wire with the receiving equipment.

7.2.2 Installation of optical workstation shall meet the following requirements:

1 The optical workstation shall be installed in machine room or equipment room;

   2 The optical workstation shall be allocated with special equipment box and the optical workstation shall be firmly installed in the special equipment box;

   3 The power supply unit of optical workstation shall be adopted AC (220V) power supply private wire, the power supply unit shall be well fixed and its spacing with the optical workstation shall not be less than 0.5 m;

   4 According to the design requirements, the optical workstation, equipment box and power supply unit shall be well grounded, and grounding terminals shall be set in the box.

7.2.3 Installation of amplifier shall meet the following requirements:

   1 The amplifier should be installed in the equipment room or weak-current room (including vertical shaft) of the building;

   2 The amplifier shall be fixed on the bottom board of the amplifier case, the indoor installation height of amplifier case should not be less than 1.2 m and the amplifier case shall be firmly installed;

   3 The amplifier case and such active equipment as amplifier shall be well grounded, and grounding terminals shall be set in the case;

   4 The input and output cables of the trunk amplifier shall be left with a surplus no less than 1m;

   5 The unused port of amplifier shall be wired in a terminal resistance of 75Ω.

7.2.4 the installation of splitter and distributor shall meet the following requirements:

   1 The installation positions and models of splitter and distributor shall be in accordance with the requirements of the design documents;

   2 The splitter and distributor shall be fixed on the bottom board of the branching distribution box;

   3 The cable shall be left with a surplus no less than one half of the box body perimeter in the splitter and distributor box;

   4 The splitter and distributor shall be connected with coaxial cables, and their connector (connector assembly) shall match the model of coaxial cable and shall be reliably connected to prevent signal disclosure;

   5 The connection between cables shall be adopted connector (connector assembly) to connect tightly without looseness or coming off;

   6 The terminals of all the branch circuits of the system and the idle output ports of distributor and splitter shall be connected with terminal resistance of 75Ω.

7.2.5 Except for being installed in the equipment room and weak-current room (including vertical shaft), on other occasions, the amplifier case, branching distribution box, pass-by case and terminal lug should be adopted with the built-in wall installation mode.

7.2.6 The wire cable laying in case body shall be in accordance with the design requirements; when being bent, its bending radius shall not be less than the specified bending radius of wire cable; each wire cable shall be reliably connected and be made with labeling.

7.2.7 The installation height of amplifier case, branching distribution box and pass-by case from the bottom edge to the ground should not be less than 0.3m.

7.2.8 In addition to those specified in Chapter 4 of this code, the laying of wire cable shall also meet the following requirements:

1 Before laying, the model and specification, route and position of the wire cables shall be checked to conform to the design drawing;

2 The minimum spacing between pipe and other pipelines shall meet those specified in Table 4.3.8 of the current professional standard GY 5078 - 2008 *Technical Regulations on Safety of CATV Distribution Network Projects*;

3 The tortuosity factor of wire cable shall not be less than the specified bending radius of wire cable and the cable shall be left with proper surplus at the turning;

4 Before laying, both ends of the wire cable shall be stuck with labels indicating the initial and end locations, and the words on the labels shall be distinct and correct;

5 During the process of laying, the wire cable shall not be squeezed, impacted or yanked, which may generate deformation.

7.2.9 The installation of coaxial cable connector shall meet the following requirements:

1 Installation of coaxial cable connector shall guarantee that the internal and external conductors of cable are reliably connected respectively;

2 Where the coaxial cable connector is connected with the equipment interface, excessive fastening shall be prevented;

3 The internal and external conductors of coaxial cable shall contact well with the stylet and shell of connector;

4 Installation of coaxial cable connector shall also meet those specified in Article 6.1.6 of the current professional standard GY 5073 - 2005 *Code for Construction and Acceptance of CATV Network Engineering*.

7.2.10 Installation of the users' indoor terminals shall meet the following requirements:

1 The terminal lug used for concealed installation shall be in accordance with the requirements of the design documents;

2 The concealedly-installed terminal lug panel shall tightly cling to the wall surface without gap around, and the terminal lug shall be installed uprightly and firmly;

3 The exposed terminal lug and panel fittings shall be complete, and the fixed screws with the wall surface shall not be less than 2 pieces.

7.3 Quality Control

7.3.1 The dominant items shall meet the following requirements:

1 The grounding of antenna system shall be separated from the grounding of lightning protection system, the equipment grounding shall be separated from the grounding of lightning protection system;

2 The feeder end of satellite antenna, the impedance matcher, the antennae arrester, the high-frequency connector and amplifier shall be firmly connected and shall be taken with rain-proof and rot-proof measures;

3 The satellite receiving antenna shall be within the protection range of lightning arrester, and the grounding resistance of antenna base shall be less than 4Ω;

4 The satellite receiving antenna shall be firmly installed.

7.3.2 General items shall meet the following requirements:

1 The installation of the equipment, devices, boxes, cases and cables of the cable television system shall meet the design requirements and shall achieve reasonable layout, tidy arrangement, fastness and reliability; the wire cables shall be exactly connected and securely crimped;

2 In the amplifier case, the equipment wiring diagram shall be stuck on the inner side of door sheet and the cable direction, signal input and output level shall be marked;

3 The panel of the concealedly installed user box shall tightly cling to the wall surface without gap around, and the installation shall be upright and secure;

4 The branch distributor and coaxial cable shall be reliably connected.

7.4 System Debugging

7.4.1 The satellite receiving antenna and system debugging shall meet the following requirements:

1 The azimuth and elevation angle of satellite receiving antenna shall be adjusted according to the received satellite parameters;

2 Signal intensity and signal quality on satellite receiver shall reach the degree position with strongest signal;

3 The grounding resistance value of antenna base shall be tested.

7.4.2 System debugging of the front end shall meet the following requirements:

1 The system debugging of front end shall be carried out after the machine room grounding system, power supply system and lightning protection system are tested to be qualified;

2 Channel of modulator shall avoid the same frequency interference field density;

3 The output level of modulator shall be adjusted to the nominal level value of this equipment.

7.4.3 System debugging of cable line and distribution network shall meet the following requirements:

1 The debugging scope shall include optical workstation, active equipment at all levels (like amplifier) and the passive equipment from cable, branch, distributor till the user terminal ; the whole debugging shall conduct positive debugging and reverse debugging;

2 Positive debugging shall measure the positive input and output technical indexes and output slope of active equipment and shall appropriately adjust such components as attenuator and equalizer to make the measured value be consistent with the design value;

3 Reverse debugging shall meet the relevant requirements of the current professional standard GY/T 180 *Technical Specifications of HFC Network Physical Upstream Transmission Path*, shall measure the reverse input and output technical indexes and output slope of active equipment and shall appropriately adjust such components as attenuator and equalizer to make the measured value be consistent with the design value; the index inspection results shall be in accordance with the requirements of the design documents.

7.5 Self-examination and Test

7.5.1 The satellite TV receiving system shall be inspected in accordance with the current professional standards GY/T 149 *Methods of Measurement for Satellite Digital Television Receive-only Earth Station—System Measurement* and GY/T 151 *Methods of Measurement for*

Satellite Digital Television Receive-only Earth Station—Outdoor Unit Measurement, and the index inspection results shall be in accordance with the requirements of the design documents.

7.5.2 Subjective assessment on system quality shall meet the relevant requirements of Section 4.2 in the current national standard GB 50200 – 94 *Technical Code for Regulation of CATV System* and the current national standard GB/T 22123 – 2008 *Methods for Picture and Audio Subjective Assessment of Digital Television Receiving Equipment*.

7.5.3 The downstream test of digital cable television system shall meet the relevant requirements of the current professional standards GY/T 106 *Technical Specification of CATV Broadcasting System* and GY/T 221 *Specifications and Methods of Measurement of Digital Cable Television System*, and the major technical requirements shall be in accordance with those specified in Table 7.5.3.

Table 7.5.3 Technical Requirements for Downstream Output Port of System

No.	Test contents		Technical requirements
1	Electric level at output port of simulated channel		60dBμV~80dBμV
2	Electric level at output port of digital channel		50dBμV~75dBμV
3	Level difference between channels	Level difference between adjacent channels	≤3dB
		Between any simulated/digital channels	≤10dB
		Level difference between simulated channel and digital channel	0dB~10dB
4	MER	64QAM, equalized closure	≥24dB
5	BER	24h, after Rs decoding (may be taken as 15 min for short-term measurement and no error code should appear)	≤1×10E-11
		Make reference to GY 5075	≤1×10E-6
6	C/N (Simulated channel)		≥43dB
7	Hum ratio of carrier wave (HUM) (simulated)		≤3%
8	Digital RF signal to noise power ratio $S_{D, RF}/N$		≥26dB(64QAM)
9	Carrier to composite second order beat ratio (C/CSO)		≥54dB
10	Carrier to composite triple beat ratio (C/CTB)		≥54dB

7.5.4 The upstream test of digital cable television system shall meet the relevant requirements of the current professional standards GY/T 180 *Technical Specifications of HFC Network Physical Upstream Transmission Path*, and the major technical requirements shall be in accordance with those specified in Table 7.5.4.

Table 7.5.4 Technical Requirements for Upstream of System

No.	Test contents	Technical requirements
1	Frequency range of upstream channel	5MHz~65MHz
2	Nominal upstream port input level	100dBμV
3	Gain inequality of upstream transmission route	≤10dB
4	Frequency response of upstream channel	≤10dB (7.4MHz~61.8MHz)
		≤1.5dB(7.4MHz~61.8MHz, within range of any 3.2 MtHz)
5	Signal hum modulation ratio	≤7%
6	Carrier wave/collection noise	≥20dB (Ra waveband)
		≥26dB (Rb and Rc wavebands)

7.5.5 Engineering construction quality of system shall meet those specified in Section 4.4 of the current standard of the nation GB 50200-94 *Technical Code for Regulation of CATV System* and in Section 2.2 of GYJ 40-89 *Acceptance and Testing Specifications for Satellite Television Earth Receiving Stations*, and the inspection on engineering construction quality shall be in accordance with those specified in Table 7.5.5.

Table 7.5.5 Inspection on Engineering Construction Quality

Item		Quality inspection
Satellite antenna	Antenna	1. Antenna support and reflector shall be firmly installed; 2. The installation orientation of antenna support shall face to the south, and the adjustable range of antenna azimuth shall meet the standard; 3. The antenna adjusting mechanism shall be flexible and continuous, the locking device shall be convenient and firm and shall be provided with measures against corrosion and dust; 4. The antenna reflector shall be provided with anti-corrosion measures
	Feed source	1. The polarization transformation structure of feed source is convenient and the performance shall not be affected during polarization transformation; 2. The horizontal polarization plane can be fine adjusted by ±45° in relation to the ground plane; 3. Feed source aperture shall be provided with sealing measures to prevent rainwater from entering into the waveguide; 4. Waterproof measures shall be taken at the joint of flanges and the plug connection place of cables
	Lightning arrester and grounding	1. The installation height of lightning arrester shall be correct; 2. The grounding wire shall meet the requirements; 3. The electrical connection at each position shall be in good condition; 4. The grounding resistance shall not be larger than 4Ω
Front machine room (including quality inspection of equipment room)		1. The ventilation, air conditioning and heat dissipation equipment etc. in machine room shall be installed in accordance with the design requirements; 2. The machine room shall be provided with lightning protection measures and grounding measures; 3. The power supply mode and power supply circuits of machine room; 4. Where standby power supply (adopting UPS power supply) is available in the machine room, the switching of backup power supplies shall be tested and the how much time of uninterrupted power supply can be guaranteed after power interruption; 5. The installation site of equipment and components shall be correct; 6. A enough length of optical cable shall be reserved in the design and shall be coiled according to an appropriate curvature radius; 7. Optical cable terminal lug shall be installed stably and be away from heat source; 8. The coupler on the single fiber cable or tail optical cable led out from the optical cable terminal box shall be inserted into socket of ODF/ODP according to the design requirements. The temporarily unused plugs and sockets shall be covered with dustproof and erosion-proof plastic caps; 9. The joints of optical fiber in terminal lug shall be securely fixed and the bending radius of the coiled surplus optical fiber in box shall be larger than the specified value; 10. The wiring shall be correct, artistic and regular; 11. The incoming and outgoing cables shall meet the requirements and the labeling shall be complete and correct
Transmission equipment		1. The model of used equipment (optical workstation / amplifier) shall be consistent with the design; 2. Each connection point shall be correct, firm and water proof; 3. The vacant end shall be treated correctly; and the enclosure shall be grounded; 4. The lightning protection measures (grounding) shall be taken and the ground resistance shall not be larger than 4 Ω; 5. Cables in box shall be lined neatly and the labeling shall be accurate and striking

Table 7.5.5(continued)

Item	Quality inspection
Branch distributor	1. The branch and distributor boxes shall be complete and the position shall be reasonable; 2. The model of installed branch distributor shall conform to the designed model; 3. The input/output connection of port shall be correct; 4. The vacant port shall be installed with terminating resistance; 5. The reserved cable length shall be appropriate and the cables in box shall be lined neatly
Cables and connectors	1. The strike, wiring and laying of cables shall be reasonable and artistic; and the labeling shall be complete and correct; 2. The bending, coiling and connecting of cables shall be meet the relevant requirements; 3. The spacing between the cable and other pipelines shall meet the requirements; 4. The specification and pattern of cable joint shall completely match the cable; 5. The cable joint shall be closely fitted to cable (crimping firmness of crimping pliers) without shedding or looseness, etc.; 6. The cable joint shall be closely fitted to the branch distributor F seat/equipment joint without looseness, etc.; 7. The joint shielding shall be in good condition without exposed shielding grid, and there is no external shield having deformation or breaking during the production process of aluminium tube cable joint; 8. After the cable joints are well fabricated, the reserved length of cable core shall be appropriate, and its length range shall be higher than the joint end face by 0~2mm; 9. The connector assembly shall be firm, waterproof and corrosion-proof
Power supply and power line	The design and construction requirements shall be met; and proper lightning protection measures shall be taken
User equipment	1. The wiring shall be regular, artistic and firm; 2. The user box shall be neatly installed at correct position; 3. The installation of user grounding box and arrester shall meet the relevant requirements

7.6 Quality Record

7.6.1 The (positive) debugging record of optical node shall be filled in Table B.0.5 of this code.

7.6.2 The (reverse) debugging record of optical node shall be filled in Table B.0.6 of this code.

7.6.3 The (positive) debugging and testing record of amplifier shall be filled in Table B.0.7 of this code.

7.6.4 The (reverse) debugging record of amplifier shall be filled in Table B.0.8 of this code.

7.6.5 The debugging record of front-end equipment shall be filled in Table B.0.9 of this code.

7.6.6 The test data record of the user electric level terminal value shall be filled in Table B.0.10 of this code.

8 Electronics Conference System

8.1 Construction Preparation

8.1.1 Technical preparation shall meet the following requirements:

1 The design documents, construction scheme, construction schedule and construction drawings of conference system shall be complete and pass the joint review;

2 Shall organize design clarification, survey the construction site, transact negotiation on technical alteration and determine the construction method;

3 The construction personnel shall be familiar with such technical documents as design scheme, construction drawings, system wiring diagram and control logic instruction as well as the relevant information;

4 The decoration of the conference place shall be inspected, the decorative materials for all parts of the room surface shall be consistent with the decoration design and shall meet the requirements for the reverberation time and background noise in the architectural acoustics design of conference system, and there shall be no such defects as echo, multi-echo and sound focusing indoors;

5 Prior to installation of equipment in control room, the decoration and sanitation shall be completed, the ground wire of antenna shall be installed and led into the indoor connecting terminal, and the incoming and outgoing wire slot shall be reserved.

8.1.2 Construction environment shall meet the following requirements:

1 All civil works of such relevant rooms as conference room, control room and transmission room have already been completed and shall meet requirements and commencement environment specified in this code;

2 The environmental requirements such as power supply, grounding, lighting, socket, temperature and humidity shall be got ready according to those specified in the design documents and shall pass the acceptance inspection;

3 The embedded concealed pipes and underground slot embedded parts required by various cables in the conference system shall be completed, the quantity, position and dimension of holes etc. shall be constructed according to design requirements and shall be accepted with qualification, and also shall be provided with accurate relevant drawings by the employer organization;

4 The ground wire of control room shall be well installed and comply with those specified in Article 16.2.1 of this code;

5 The construction site shall have the approaching conditions and be able to guarantee the construction safety and safety utilization of electric power.

8.2 Equipment Installation

8.2.1 The setting of cabinets shall meet the following requirements:

1 The cabinet shall be installed on the cabinet underframe and should not be directly placed on anti-electric floor, and the underframe shall be firmly connected with the ground;

2 Arrangement of cabinets shall be reserved with the spacing for maintenance, the clear distance between the machine surface and the wall shall not be less than 1.5 m, clear distance between the machine back and side and the wall (where maintenance is required) shall not be less than 0.8 m; if the cabinets are arranged in front and back, the clear distance of arrangement between cabinets shall not be less than 1 m;

3 The horizontal installation position of cabinet shall meet the construction drawing design and its deviation shall not be larger than 10 mm; the vertical deviation of cabinet shall not be larger than 3 mm;

4 Where several cabinets are installed in order, the front faces of every row of cabinets shall be on the same plane, and the adjacent cabinets shall be close to each other tightly;

5 Various modules on cabinet shall be installed firmly without buckling damage, and the finish, if sheds, shall be repaired; the modules, if any damaged, shall be repaired or replaced;

6 Marks indicating the equipment name or function shall be set on cabinet, and the marks shall be correct, distinct and complete.

8.2.2 The power supply and grounding of equipment shall meet the following requirements:

1 The conference system shall be set with dedicated shunt switchboard, the capacity of each line shall be determined in accordance with the actual conditions and a certain surplus capacity shall be reserved;

2 The audio and video equipment of conference system shall be adopted the power supply of the same phase;

3 All the metal enclosures of equipment, the metallic pipelines, metallic wire slots and metal structure of building in control room shall be carried out with equipotential bonding and also shall be grounded;

4 Power supply circuit of conference system should be adopted the power supply circuit with low interference on the incoming household end of building, the protective earth wire (PE wire) shall be separated from the zero conductor of AC power supply, and the unbalanced current of zero conductor shall be prevented from generating severe interference to the conference system; and the noise interference voltage of protective earth wire shall not be larger than 25 mV;

5 The power supply for the lighting equipment (including dimming equipment) of conference room and the equipment of audio and video systems of conference place should be adopted the shunt feeding mode;

6 Control room should be taken with anti-electrostatic measures, and the electrostatic grounding may share the working grounding of the system;

7 When laying wire cables, the sheath, shielding layer and core wire shall be free from damage and rupture phenomenon and shall be made with obvious labeling.

8.2.3 In addition to those specified in Chapter 4 of this code, the pipeline laying also shall meet the following requirements:

1 After the pipeline in suspended ceiling entering into the control room, the pipeline shall be perpendicularly led into the antistatic floor along wall surface at a nearby place and then shall be led into the wire slot in the bottom of cabinet along the ground;

2 The ground pipeline shall be led into the electrostatic floor in control room by clinging to the ground and then shall be led into the metallic wire slot in the bottom of cabinet;

3 The signal line and the strong-current wire pipes shall be laid separately by adopting metallic pipes;

4 The underground wire slot from cabinet to console shall be laid under the antistatic floor of control room;

5 When installing the single-side or double side cable conduits along wall, the equipment support frames built in wall shall be firm and reliable, and the spacing between supporting points shall be uniform, regular and consistent.

8.2.4 Installation of conference speech system shall meet the following requirements:

1 For the professional wired conference system adopting serial connection mode, the connecting wire cable between microphones shall be firmly terminated;

2 For the system composed by directly connecting the microphones to sound reinforcement facility, the microphone transmission line shall adopt the dedicated shielded line;

3 Where mobile microphones are adopted, the wire cable protection shall be done well and the wire cables shall be protected against damages;

4 Where the transmission distance of wireless microphones is relatively remote, outboard receiving antenna shall be additionally installed; if the outboard receiving antenna is installed on table top, the fixed holder should be equipped.

8.2.5 Installation of loudspeaker system shall meet the following requirements:

1 Installation of loudspeaker system shall be consistent with the design, the centralized, dispersed installation mode or the combination of both may be selected and shall meet the requirements for whole site coverage and sound field uniformity;

2 The loudspeaker system shall be fixed safely and reliably, its installation height and installation angle shall meet the requirements of sound field design;

3 Where the building structure is used to install such accessories as support or hanger rod of the loudspeaker system, the bearing capacity of this building structure shall be inspected;

4 Where the loadspeaker system is concealedly installed, the dimension of the concealed installation space shall be large enough (this space shall be carried out with sound absorption treatment) to ensure that the speaker is able to have radiation angle adjustment in this space; the acoustic permeability of the speaker shield shall meet the requirements; if the shield is cell structure, its scantling (width and depth) should not be larger than 20 mm;

5 Where the loadspeaker system is installed on the ceiling, the loadspeaker placement shall meet the requirements for sound field uniformity and artistic layout;

6 The loadspeaker system shall be far away from microphones, the axial pointing direction shall not be directed to microphones and shall avoid causing self-excited squeal;

7 LoadSpeaker system shall be taken with reliable safety precautions and shall not generate mechanical noise at working;

8 When carrying out with hoisting installation of the loadspeaker box and horn speaker, original accessory hanger fixtures shall be adopted; where the original parts are unavailable, the dedicated hanging fixtures for loadspeaker box such as steel wire rope or galvanized iron chain may be selected;

9 The outdoor loadspeaker system shall be with moistureproof and anticorrosive property, and the fasteners shall be with adequate bearing capacity;

10 Loadspeakers used in areas with fire hazard shall be made from flame retardant materials or be adopted with flame-retardant back cover; the broadcasting loadspeakers shall be able to work normally for short term under sprinkling conditions.

8.2.6 Installation of audio equipment shall meet the following requirements:

1 Equipment installation sequence shall be consistent with the signal flow;

2 The cabinet installation sequence shall be with light one above and heavy one below, the equipment such as wireless microphone receiver shall be installed at upper part of cabinet; the heavy equipment such as power amplifier shall be installed at the lower part of cabinet and also shall be supported by guide rail;

3 All wire cables of the system shall be led into the elevated floor in control room through metallic conduits, wire slots and then shall be led to place below cabinet and console;

4 In reserved power supply box in control room, measures against electromagnetic impulse shall be taken and the stabilized voltage supply device with wave filtering function shall be allocated; the electricity supply capacity shall meet the capacity when all system equipment are switched on; if the system possesses fire emergency broadcast function, electricity shall be supplied according to Class I load; mutual mapping of double power supply end shall be adopted and uninterruptible power supply shall be arranged;

5 The sound console should be installed on the operating desk for operating and regulating by tuning personnel; equipment such as program source that are frequently operated shall be installed on the position where is easy to operate;

6 Cabinet shall be fixed onto the foundation section steel by adopting bolts, and the perpendicularity of the installed cabinet shall be inspected and adjusted; the console shall firmly fixed with the foundation and shall be placed regularly;

7 The equipment in cabinet shall be installed stably and regularly, the panels shall be arranged neatly, the panel screws shall be tightened up; the equipment with rails shall be able to be pushed and pulled flexibly; the internal wire cables shall be lined neatly in categories; sufficient heat dissipation gap shall be left between equipment to install the ventilation panel or blind plate;

8 The connectors at both ends of cable shall be selected with the qualified products and shall be fabricated by adopting special tools while cold joint or rosin joint is disallowed; the position where connector needs to be crimped shall guarantee the crimping quality, and looseness and shedding are disallowed; the connectors shall be strictly inspected after being fabricated, and only the qualified one may be used; the balanced wiring mode shall not be disturbed by external electromagnetic field and shall be with good acoustic quality;

9 Labeling indicating end property and usage shall be set nearby connectors at both ends of the cable; misconnection and lack of connection are disallowed;

10 The time sequence power supplies shall be connected in turn according to the starting sequence and their installation positions shall give attention to the length of equipment power lines;

11 Corresponding shock eliminators shall be selected in accordance with equipment and materials in the cabinet.

8.2.7 Installation of video equipment shall meet the following requirements:

1 When installing the display screen, phenomena such as reflected light and glare shall be avoided; walls and floors should be used material which is uneasy to reflect light;

2 Where the distance of transmission cable exceeds the standard length supported by the selected port, signal amplifier and line compensation equipment shall be used or the optical cables shall be selected for transmission;

3 Display equipment should be supplied with electricity separately by using wave filtering socket of power supply;

4 Display shall be firmly installed, the load bearing capacity of the wall and support for fixing the equipment shall meet the design requirements; appropriate installation support frame, hanger and fasteners shall be selected, and the screws and bolts shall be fastened in place;

5 The installation positions of the large screen display embedded in wall, the wall-mounted display and the like shall meet the requirement for optimal sighting distance.

8.2.8 Installation of simultaneous interpretation equipment shall meet the following requirements:

1 Where wired simultaneous interpretation system is adopted, the listening devices such as headset jacks, volume controls and shunt option switches shall be set at the audience seats;

2 Where wireless simultaneous interpretation system is adopted, the quantity of and installation positions of wireless transmitters shall be exactly determined according to the seat arrangement and in combination with the effective wireless coverage range;

3 For simultaneous interpretation, private interpreter room should be set and shall meet the following requirements:

    1) The interpreter room should be set with sound-proof observation window and shall have the condition to observe the chairman platform scene;

    2) Outside the interpreter room, transliteration work indicating lamp or board shall be set;

    3) Interpreter room may be adopted the stationary type or mobile type.

8.2.9 Installation of video conference equipment shall meet the following requirements:

1 Video conference system shall include such parts as video conference multipoint control unit, conference terminal, access gateway, audio sound reinforcement and video display;

2 Microphone arrangement should be kept clear from the primary radiation area of loadspeaker and shall reach such requirements as uniform natural and distinct sound field and good feeling of sound source;

3 Video camera arrangement shall make the shot object be in the angular field of view, should take pictures from multiple directions and shall be able to obtain full view or local close-up view of the conference place;

4 Arrangement of monitor or large display screen should make the participants be within preferable sighting distance and angular field of view;

5 The lighting conditions at the video signal acquisition area in conference place shall meet the following requirements:

    1) Colour temperature of light source: 3,200K;

    2) Average illuminance at the chairman platform area should be 500 lx~800 lx, the average illuminance of general areas should be 500 lx and that at the projection TV screen area should be less than 80 lx.

8.3 Quality Control

8.3.1 The dominant items shall meet the following requirements:

1 Installation levelness of equipment in cabinet shall be guaranteed and the construction shall not be carried out in dusty or unclean environment;

2 Equipment shall be firmly installed;

3 Length of signal cable shall not exceed the design requirement;

4 Video conference shall be possessed of higher speech intelligibility and appropriate reverberation time; where the volume of conference place is less than 200m^3, the reverberation time should be 0.4s~0.6s; where the video conference room is also used for other functions, the reverberation time should not be larger than 0.6s; where the volume of conference place is larger than 500m^3, it shall comply with the current national standard GB/T 50356 *Code for Architectural Acoustical Design of Theater Cinema and Multi-use Auditorium*.

8.3.2 General items shall meet the following requirements:

1 Prior to cable laying, the integral path inspection shall be conducted;

2 Prior to equipment installation, the equipment shall be pre-checked by powering on, and the faulted equipment shall be treated timely.

8.4 System Debugging

8.4.1 Prior to system debugging, the fabrication of field equipment wiring diagram and control logic instruction shall be completed.

8.4.2 The debugging preparation shall meet the following requirements:

1 Ground resistance shall be inspected; if it does not conform to the design requirements, debugging by powering on shall not be started;

2 The technical personnel shall be familiar with the control logic and prepare the debugging record sheet;

3 Prior to system debugging, every equipment shall be confirmed of having no quality problem and then electrifying may be started;

4 The model and installation position of various equipment shall meet the design requirements;

5 The service power supply voltage marked on various equipment shall be correspond to the power supply voltage at the service site;

6 Specification and model of wire cables connecting the equipment shall be inspected, and the connection of wire cables shall be correct without looseness or cold solder joint;

7 Before powering on, the switches and knobs of every equipment shall be set to the original position.

8.4.3 Debugging of audio equipment shall meet the following requirements:

1 Power supplies of corresponding equipment shall be switched on according to different functions of conference system and the equipment shall be confirmed of normal working;

2 It shall be confirmed that the relevant equipment and database of record system are operating normally;

3 It shall be confirmed that the equipment of the system are working normally, and the equipment parameters shall be adjusted;

4 It shall be confirmed that the system is operating normally, and the system shall be carried out with fine adjustment according to design functional requirements to reach the optimal integral effect;

 5 Objective measurement index shall reach the requirement of speech intelligibility STPA;

 6 The system index shall meet the requirements for acoustic characteristic index of sound reinforcement system stated in the current national standard GB 50371 *Code for Sound Reinforcement System Design of Auditorium*;

 7 Subjective audition of system after being debugged shall reach distinct language, full music and uniform sound field.

8.4.4 Debugging of video equipment shall meet the following requirements:

 1 The power supply of video equipment shall be switched on, the video signals and computer generated signals shall be respectively accessed into the display equipment, the image quality shall meet the relevant requirements of the current national standard GB 50348 *Technical Code for Engineering of Security and Protection System*;

 2 The projector shall be adjusted according to the position of curtain and shall be positioned after being debugged to appropriate position; the projection focus and gradient shall be adjusted till the image is distinct and upright;

 3 Camera of conference speech system shall be able to automatically follow the speaker and shall automatically focus and amplify; the interlocking video display units shall display the images of speaker;

 4 Conference information processing system may realize the rapid switching among the multiple channels of video signals and data signals through matrix; and the images shall be stable and reliable;

 5 The conference record system shall be able to store the scenes at the conference place and invoke and play them at will;

 6 After the system is debugged, its image resolution, continuity, tone and color saturation shall reach the design index requirements.

8.4.5 Debugging of conference units shall meet the following requirements:

 1 Prior to electrifying, the switches and knobs of every equipment shall be set to the required positions; the software installation and the parameter setting and adjustment shall be completed according to the equipment requirements;

 2 The equipment shall be preheated when being powered on for the first time; if no abnormal phenomenon is observed, the normal operation may be carried out;

 3 It shall be confirmed that the communication of conference units with the host is in good condition and the functions and operation are normal; the language sound enlargement of every conference unit shall be clear;

 4 Functions of the conference units shall be inspected in accordance with the equipment operation instruction and design documents.

8.4.6 Debugging of video conference system shall meet the following requirements:

 1 The image resolution and image frame rate shall meet the relevant standards of China;

 2 The sound shall be clear and continuous without noise or echo.

8.4.7 Debugging of simultaneous interpretation system shall meet the following requirements:

 1 The system shall have the function of automatically switching to the field language; where the field speech uses the same language as the interpreter, the microphone of interpretation device shall be switched off, the interpretation control host shall automatically switch this interpretation channel to the field language;

2 Call and technical support functions. Every interpretation bench shall be provided with the autonomous channel to call the chairman and technical personnel;

3 Interpretation channel locking function. The system shall be set with the indicator lamp showing the occupation condition of channels, and different interpretation languages shall be avoided occupying one channel;

4 Independent voice monitoring function. The interpretation control host may monitor all the channels and the field languages and shall be with independent sound volume control function.

8.4.8 Debugging of central control equipment shall meet the following requirements:

1 Control software shall be prepared according to the control logic diagram, and the control effectiveness of equipment shall be tested one by one; the equipment shall have the function of using various wired or wireless touch screens to realize the remote control of audio, video, lighting, curtain as well as environment at conference place, and the debugging record shall be filled;

2 After being debugged, the central control system shall have the following functions:

 1) Volume control function;

 2) Shall have normal connection and communication with the conference discussion system, and shall control the free switching and distributing of audio and video;

 3) Shall be able to control the serial port equipment by multichannel RS-232 control port;

 4) Shall control equipment such as DVD and TV set by infrared remote control;

 5) Shall control equipment such as electric projection screen, electric curtain and projector lift by multichannel digital I/O control port and light-current relay control port;

 6) Shall be able to expand to connect such peripheral equipment as multiple power supply control units, light control unit, wireless transceiver and wall-mounted panel.

3 The system shall have the user-defined scene storage and scene invocation functions;

4 Shall realize the intelligent management and operation of systems in the conference place by the central control system.

8.5 Self-examination and Test

8.5.1 Inspection on audio sound enlargement, simultaneous interpretation and vote record functions shall meet the following requirements:

1 Shall be able to play multichannel audio signals;

2 The played music shall be with distinctive nuance, full sound and adequate sound pressure level;

3 Wired microphones and conference microphones shall be used normally;

4 During subjective audition of language sound reinforcement, there shall be no squeal, the sound shall be clear and the sound pressure level shall be adequate;

5 During subjective audition of human vocal singing, there shall be no squeal, the sound shall be clear and the sound pressure level shall be adequate;

6 Objective measurement indexes shall reach the requirements for speech intelligibility STI and the requirements for corresponding acoustic characteristic design indexes;

7 Positions of auditoriums shall be without obvious audible background noise;

8 Precision rate of vote record shall be as high as 100%.

8.5.2 Inspection on video and audio switching and display systems shall meet the following requirements:

1 Shall be able to display different kinds of image signals required by the design on various display equipment;

2 The image signal shall be clear and stable and without dithering or flickering.

8.5.3 Inspection on centralized control system shall meet the following requirements:

1 Shall be able to control the switching of different kinds of image signals on various display equipment;

2 Shall be able to control the switching of audio signals;

3 Shall be able to control the sound volume level and the rapid conversion among various operation modes;

4 Shall be able to control the pattern switching of display system and the invocation of various images;

5 Shall be able to control the dimming, switch and mode selection of lighting system;

6 Shall be able to control the switch and the operation of all functions of electrical equipment.

8.6 Quality Record

8.6.1 In addition to complying with those specified in Section 3.7 of this code, the quality record of conference system also shall comply with the relevant requirements of national or professional standards.

9 Broadcasting System

9.1 Construction Preparation

9.1.1 Preparation for materials and equipment shall not only meet the requirements of Article 3.3.2 of this code, but also meet the following requirements:

1 The specification, model, quantity of equipment shall meet the design requirements, the product shall be accompanied with quality certificate and be marked with "CCC" (China Compulsory Product Certification);

2 All active components shall be subjected to power-on inspection, and confirmed their actual functions and technical indexes shall be consistent with nominal ones;

3 For hardware devices and materials, keep focus on inspecting such items as the safety, reliability and electromagnetic compatibility.

9.2 Equipment Installation

9.2.1 Bridge and pipeline laying shall not only meet the requirements specified in Chapter 4 of this code, but also meet the following requirements:

1 The outdoor broadcasting transmission cables shall be passed through pipes which are buried underground or laid in cable trenches; indoor broadcast transmission cables shall be passed through pipes or laid in wire slots;

2 Cables used for power transfer of the broadcasting system shall be laid in exclusive wire slots and wire pipes;

3 Where the broadcasting system has the function of fire emergency broadcast, they shall be laid in flame retardant wire slots, wire pipes and wire cables;

4 For the power transfer line of broadcasting system, its insulation voltage grade shall be compatible to its rated transmission voltage, its joints shall not be naked; for joints with unequal potential, they shall be subjected to insulating treatment respectively;

5 The transmission cable for broadcasting system shall be possessed of fewer joints which shall be well wrapped and placed in the inspection box.

9.2.2 The installation of loadspeaker shall meet the following requirements:

1 The height, horizontal and vertical pointing direction of the loadspeaker shall be determined by the sound field design and the field situation, and shall meet the following requirements:

    1) The acoustic radiation of the loadspeaker shall be pointed to broadcasting service area;

    2) Where there are high buildings and topography, around, echo caused by improper installation shall be avoided;

2 The joints between the loadspeaker and broadcast line shall be well contacted, and the joints with different potential shall be insulated respectively; and they should be connected by using crimping sleeves and crimping tools;

3 The loadspeaker shall be safely and reliably installed and fixed. These facilities used for

installing the loadspeaker, such as the lamppost, truss, wall, shed roof and fasteners, shall be possessed of sufficient bearing capacity;

　　4 The loadspeakers installed outdoor shall be protected against moisture, rain and mould. When they are installed at the contaminated zone with salt mist, sulfide, etc., anti-corrosion measures shall be taken.

9.2.3 Except the loadspeaker, other equipment should be installed on the console, equipment cabinet or rack which located in the monitoring room (or machine room); where there is no monitoring room (machine room), the console, equipment cabinet or rack shall be installed at the place where is safe and convenient to control.

9.2.4 The arrangement of equipment in the equipment cabinet and rack shall allow the operator on duty to see them clearly for the frontage of a majority of the equipment, and allow the operator to operate and adjust them rapidly and conveniently as well as to monitor the running signals of each equipment.

9.3 Quality Control

9.3.1 The dominant item shall not only comply with the relevant requirements of Article 4.2.10 in the current national standard GB 50339 – 2003 *Code for Acceptance of Quality of Intelligent Building Systems* but also meet the following requirements:

　　1 The installation of equipment, such as the loadspeaker, controller and socket, shall be firmly and reliably; the connection of wire shall be in alignment; the wire number shall be correct and distinct;

　　2 **Where the broadcasting system has the function of emergency broadcast, the emergency broadcast shall be controlled by the fire-fighting extension set, and shall be possessed of the highest priority; where fire and emergency occurring, the system shall be able to be switched to the emergency broadcast for broadcasting at the maximum sound. The broadcasting system shall be able to broadcast caution signals (including alarm whistle), alarm voice or real-time command voice to the relevant broadcast domain within 10s after it has been triggered by manual operation or alarm signals. Take the noise of on-site environment as the reference and the SNR (Signal to Noise Ratio) of emergency broadcast shall not be less than 15dB.**

9.3.2 Quality control for general item shall meet the following requirements:

　　1 Ceiling loadspeakers in the same room shall be arranged uniformly. Elevation of loadspeaker enclosure, control unit and socket shall be consistent, flat, smooth and firm; the surrounding of the loadspeaker shall be free from crevasse; the decorating shield shall be free from damage and shall be flat and smooth;

　　2 The wire connection of each equipment shall be correct, reliable and firm; cables (wire) in chamber shall be in alignment; wire serial number shall be correct and distinct. Where there are lots of lines, these lines shall be bundled, and proper space shall be left for them in the case (box).

9.4 System Debugging

9.4.1 Debugging preparation shall meet the following requirements:

　　1 The equipment interface of the broadcasting system and the third party linkage system shall be completed and shall meet the design requirements;

2 Various selective switches of the equipment shall be placed at the designated position;

3 Prior to energizing the equipment, the output voltage of all power supply transformers shall be inspected and shall meet the requirements of the equipment specification;

4 All levels of hardware equipment shall be power on step-by-step, and shall pass the self-examination according to operation sequence of the equipment specification;

5 Debugging materials, including the system network structure drawing, equipment wiring drawing and equipment operation, installation and maintenance instructions, shall be complete.

9.4.2 Equipment debugging shall meet the following requirements:

1 When debugging with power on, knobs of all equipment shall be adjusted to the minimum position, and the machine shall be started with power on step by step in accordance with the order from preceding stage to backward stage;

2 All audio inputs shall be adjusted to the appropriate situation; acoustic audition shall be conducted towards each broadcast domain; conduct preliminary debugging according to the inspection result;

3 After the installation of loadspeaker has been completed, all broadcast domains shall be subjected to detection and audition one by one;

4 The function of each broadcast domain and the overall system shall be inspected, and adjustment shall be carried out according to the inspection result to ensure the emergency functions of the system meet the design requirements;

5 Normal running and operation shall be simulated designedly and repeatedly and the operation result shall meet the design requirements;

6 Persistent power on time for system debugging shall not be less than 24h;

7 The electroacoustic performance index of the system shall be tested, and then adjusted. The electroacoustic performance index of the system shall meet the design requirements;

8 System debugging shall be well recorded.

9.5 Self-examination and Test

9.5.1 Inspection of transmission line shall meet the following requirements:

1 Each transmission wiring shall be correct and shall be free from such faults as short circuit, open circuit and wiring cross;

2 The serial number of the terminals shall be complete and correct.

9.5.2 Measurement of the insulation resistance shall meet the following requirements:

1 The insulation resistance between lines, and between the line and ground shall be measured;

2 Resistances of each return circuit shall be measured over the shunt and return circuit;

3 The insulation resistance between broadcasting lines shall not be less than $1M\Omega$.

9.5.3 Measurement of the grounding resistance shall meet the following requirements:

1 The power frequency grounding resistance of broadcasting power amplifier and arrester shall not be greater than 4Ω;

2 The ground resistance of public grounding system shall not be larger than 1Ω.

9.5.4 Power supply test shall meet the following requirements:

1 Make-break operation shall be conducted over the power switch to inspect the power supply display signal;

    2 The shifter of the standby power supply shall be tested for inspecting the output voltage of the storage battery;

    3 The rectifying charging unit shall be inspected and measured;

    4 Power off simulation test shall be conducted.

9.6 Quality Record

9.6.1 Table B.0.11 in this code shall be filled for the Measurement Records for Project Electroacoustic Performance of broadcasting System.

10 Information Facilities System

10.1 General Requirements

10.1.1 The information facility and system shall cover the communication access system, telephone switching system, information network system, generic cabling system, indoor mobile communication coverage system, satellite communication system, cable television system, broadcasting system, electronics conference system, clock system, information guidance and release system, calling system, ticketing and checking system and other relevant information communication system.

10.1.2 The construction of generic cabling system shall meet the requirements of Chapter 5 in this code; the construction of information network system shall meet the requirements of Chapter 6 in this code; the construction of cable television system shall meet the requirements of Chapter 7 in this code; the construction of electronics conference system shall meet the requirements of Chapter 8 in this code; the construction of broadcasting system shall meet the requirements of Chapter 9 in this code.

10.1.3 The construction of indoor mobile communication coverage system shall meet the requirements of the current professional standards YD/T 5120 *Specification on Indoor Coverage Engineering Design for Wireless Communication System* and YDJ 31 *Technical Code for Installing and Acceptance of Communication Power Equipment*.

10.1.4 The construction of satellite communication system shall meet the relevant requirements of the current professional standards YD/T 5050 *Specifications of Engineering Design for the Domestic Satellite Communication Earth Station*, YD/T 5028 *Specifications of Engineering Design for Very Small Earth Station (VSAT) System of Domestic Satellite Communication* and YD 5017 *Specification of Execution and Acceptance of Equipment Installation Engineering for Domestic Satellite Communication Earth Station*.

10.2 Equipment Installation

10.2.1 Equipment Installation of telephone switching system and communication access system shall meet the following requirements:

    1 Before installing the telephone switching equipment, environmental conditions in the machine room shall be inspected and the environmental conditions shall meet the relevant requirements of Chapter 14 in current professional standard YD/T 5076-2005 *Design Specifications for PSTN Exchange Installation Engineering*;

    2 The cabinet of the switchboard shall be installed according to the engineering design plan, and the vertical misalignment at the upper and lower ends shall not be greater than 3mm;

    3 The connectors inside the cabinet of the switchboard shall be firmly connected with the rack;

    4 The cabinets shall be arranged in line and the deviation in each 5m shall not be greater than 5mm;

5 The installation position of the cabinet shall be correct and the cabinet row shall be in order, the adjacent two cabinets shall be tightly closed and the connecting place between two cabinet surfaces shall be free from obvious unevenness;

6 The installation location of the master distribution frame shall meet the design requirements;

7 The vertical misalignment at the upper and lower ends of each kind of distribution frame in line shall not be greater than 3mm, and the horizontal deviation of the pedestal shall not be greater than 2mm per meter;

8 All kinds of words and symbols shall be correct, clear and complete;

9 The terminal equipment shall be completely arranged and well installed, and the marks shall be complete and correct. ;

10 The rack and distribution frame shall be strengthened according to the aseismic requirements of the construction drawing;

11 For direct current power line and the power line in the cabinet row which connects with it, the insulation resistance between the positive line and the negative line and between the negative line and the ground shall be tested, and all of them shall not be less than 1MΩ;

12 The insulation resistance among the core wires of AC power line used for the switching system and the insulation resistance between the core wire and the grounding wire shall not be less than 1MΩ;

13 The AC power line used in the switching system shall be provided with protective grounding wire;

14 Before the switchboard is power on, the following contents shall be inspected:

1) The quantity, specification and wiring of all kinds of circuit boards and the installation location of the rack shall be consistent with the construction drawing design document, and all identifications shall be correct and complete;

2) The specification of the fuse protector possessed by the rack shall meet the requirements, when inspecting, power switch of each functional unit shall be in the closed position;

3) All kinds of selective switches of the equipment shall be in the initial position;

4) The specification of the power line of the equipment and the grounding wire shall meet the design requirements and the terminating shall be correct and firm;

15 The master power input voltage in the machine room shall be measured and when it is confirmed to be normal, it can be tested with power.

10.2.2 Equipment installation for the clock system shall meet the following requirements:

1 The central master clock, time server, supervisory computer and shunt output interface box shall be installed in cabinet of the machine room, and shall meet the following requirements:

1) The shunt interfaces and slave clock shall be connected according to the design and equipment installation drawing;

2) The distance between installation location of the central master clock and the GPS antenna should not be greater than 300m;

3) The installation of time server and supervisory control computer shall comply with those specified in Article 6.2.1 and Article 6.2.2 of this code.

2 The slave clock shall be installed firmly; the mounting height of the wall-mounted slave clock should be 2.3m~2.7m; the mounting height of the hanging slave clock should be 2.1m~2.7m;

 3 The antenna shall be installed at outdoor, and at least shall be free from sheltering with three sides, and shall be placed in the lightning protection zone of the building;

 4 The antenna shall be fixed on the metal base on the wall surface or on the roof;

 5 The installation of large-scale outdoor clocks shall meet the following requirements:

  1) The influence of wind shall be considered according to the dimension of the outdoor clock, and the outdoor clock should be equipped with a supporting bracket;

  2) For buildings of steel structure, the supporting bracket of the outdoor clock shall be installed by welding;

  3) For buildings of concrete structure, the supporting bracket of the outdoor clock shall be installed by the way of embedded steel frame;

  4) Lightning protection device shall be installed according to the design requirements;

  5) Leakage proof and rainproof sealing measures shall be well taken.

10.2.3 The installation of the information guidance and release system shall meet the following requirements:

 1 The system server and the workstation shall be installed in the cabinet of the machine room and shall meet the requirements of Chapter 6 in this code;

 2 The installation location of the touch screen and the display screen shall make no influence towards the pedestrian passageway;

 3 The touch screen and the display screen shall be installed at dry places and at places where without strong electromagnetic radiation source;

 4 The bearing capacity of the steel frame for installing the landing type display screen shall be determined on site by coordinating with relevant professions and shall meet the design requirements;

 5 The display screen installed outdoor shall be well taken with anticreep and rain-proof measures, and shall meet the protection grade standard of IP 65.

10.2.4 The installation of the call-intercom system shall meet the following requirements:

 1 The installation of call-intercom system used for the hospital shall meet the following requirements:

  1) The mounting height of the wall-mounted host machine should be 1.2m~1.8m;

  2) The desk type host machine should be installed in front of the office desk of the operator on duty, and the installation location of the signal concentrator shall close to the host machine;

  3) The call button should be installed at the location where is convenient to touch;

  4) The pulling type challenge switch may be installed at the place where has no influence on visual effect and where is easy to pull the line as appropriate;

  5) Near the installation location of the wireless paging antenna, there shall be free from strong electromagnetic radiation source;

 2 The installation of the call-intercom system for the neighborhood buildings shall meet the following requirements:

  1) The mounting height of the outdoor call-intercom terminal should be greater than 1.2m;

2) The outdoor call-intercom terminal shall be well taken with anticreep and rain-proof measures;

3) The installation location of the signal concentrator shall be close to the host call machine.

10.2.5 The installation of the ticketing and checking system shall meet the following provisions:

1 All host machines for the ticketing and checking system shall be well grounded and shall be safe and reliable;

2 The installation of the ticket gate machine shall meet the following requirements:

1) The installation shall meet the design requirements;

2) The power cable and the communication transmission cable of the ticket gate machine and shall be laid in concealed pipes. The coupling end shall be adopted dedicated connecting device;

3) Each ticket gate machine shall be possessed of protective measures for anticreep;

3 The ticket selling machine shall be installed firmly.

10.3 Quality Control

10.3.1 The dominant items shall meet the following requirements:

1 The test stage, test content and method and performance index of the telephone switching system and communication access system shall meet the requirements of the current professional standard YD 5077 *Accepting Specifications for SPC Exchange Installation Engineering*;

2 The transmission rate, signal mode, physical interface and interface protocol of the communication system to connect the channel of common user network shall meet the design requirements;

3 The time control of temporal information equipment, master clock and slave clocks of the clock system must be accurate and synchronous;

4 The multi-media display screen must be installed firmly. The power supply and communication transmission system must be reliably connected to meet the requirements for application;

5 The call-intercom system shall timely and correctly response to the calling, and the image and voice shall be distinct;

6 The statistics of ticketing data and checking data for the database management system of ticketing and checking system shall be accurate;

7 The automatic passageway gate machine of the ticketing and checking system must be able to correctly response and reliably operate.

10.3.2 General items shall meet the following requirements:

1 The identifications of equipment and wire cable shall be distinct and clear;

2 When installing various business boards and cable of business boards, the signal wire and power source shall be led into respectively;

3 Installation of each equipment, device, box, case, wire cable, etc. shall meet the design requirements, and they shall be of reasonable arrangement, in alignment, firm and reliable, with correctly connected wire cables and with firm compression joints;

4 The feeder connector shall be installed firmly and the its contact shall be in good condition. The connector shall be taken with rain proof and anti-corrosion measures.

10.4 System Debugging

10.4.1 The debugging preparation shall meet the following requirements:

1 Prior to system debugging, the debugging plan and test plan shall be formulated and be approved in the joint hearing;

2 The specification and installation of the equipment shall meet the design requirements; the installation shall be stable and the enclosure shall not be damaged;

3 The power cables shall be measured by a 500 V tramegger, and the insulation resistance between its cable cares, cable core and the ground wire shall not be less than 1 MΩ;

4 Identifications of the equipment and wire cable shall be complete and accurate, and shall meet the design requirements and the requirements of Chapter 5 in this code;

5 The equipment cabinet, control box, support, equipment as well as the shielded wire cable requires to be grounded and the coaxial cable shall be well grounded;

6 The voltage and power of the power supply and distribution of each system shall meet the design requirements.

10.4.2 Debugging of the information facility and system shall meet the following requirements:

1 Equipment in each system shall be able to make response to orders of the system software in time;

2 During system debugging, the working state and operating log of the software shall be recorded and inspected in time, and errors shall be able to be corrected;

3 During system debugging, the response state of the system equipment against the commands of the system software shall be recorded and inspected in time, and errors shall be able to be corrected;

4 The performance test may be conducted only after the functional test;

5 Where operation errors occurring in the process of debugging, or the system function or performance cannot meet the design requirements, the problem reporting table for system debugging shall be filled in, and treatment shall be carried out and recorded in time.

10.4.3 The debugging and test of the telephone switching system shall meet the following requirements:

1 Power up the equipment grade by grade. After the equipment is powered on, inspect all racks and the output voltage supplying power for the equipment shall meet the design requirements;

2 The telephone switching system shall be normal upon self-examination, and the clock synchronization, clock grade and performance parameters shall meet the design requirements;

3 Debugging of the service system, online billing system and switching centralized monitoring system of the telephone switchboard shall be free from system failure, and shall be provided with corresponding test reports.

10.4.4 The debugging and test of the communication access system shall meet the following requirements:

1 Power up the equipment grade by grade. After the equipment is powered on, inspect all racks and the output voltage supplying power for the equipment shall meet the design requirements;

2 The installation of the system and the equipment installation shall meet the design requirements.

10.4.5 The debugging and test of the clock system shall meet the following requirements:

1 The software system parameters, processing function and communication function of the allocated server and computer, shall meet the design requirements;

2 The fault equipment and software shall be repaired or replaced;

3 The configuration management, performance management and fault management shall be carried out for the master clock, slave clock and time server in the system through the supervisory control computer;

4 The slave clock shall be subjected to such functional debugging as time adjustment, time-tracing and stop through the supervisory control computer, and networking connection and control shall be realized for all clocks;

5 The synchronization between the master clock and the timing signal receiver, synchronization of the master clock over the slave clock shall be debugged, and all clocks shall be synchronous with GPS;

6 Functions of switching between the master and the salve and the automatic recovery function of the dual-master clock system shall be debugged;

7 All equipment shall be subjected to uninterrupted functional and performance tests and shall meet the following requirements:

　　1) During the test, systematic fault or reliability fault of the clock shall not occur, the time shall be accurate; otherwise, test shall be done again after the clock has been repaired or replaced;

　　2) The test process, repairing measures and test results shall be recorded;

8 After passing the test, functional and joint debugging tests between the equipment and other system interfaces shall be carried out, and shall meet the following requirements:

　　1) The interfaces between the clock system and other systems shall be correct;

　　2) The clock system shall provide the reference time to other subsystems according to the design requirements.

10.4.6 The debugging and test of the information guidance and release system shall meet the following requirements:

1 The software system parameters, processing function and communication function of the allocated server and supervisory control computer shall meet the design requirements;

2 Single-machine debugging shall be carried out over display equipment of the system to make each display screen achieve correct luminance and color display;

3 Loading contents of words and image, debug and detect each terminal which shall correctly display the released content;

4 Debug and detect each function of the software, and it shall meet the design requirements;

5 Test the sound and video telecasting quality of the terminal, and both of them shall be qualified;

6 All equipment shall be subjected to 24h uninterrupted functional and performance tests after system debugging, and shall meet the following requirements:

　　1) During the test, systematic fault or reliability fault shall not occur, no blind spot shall be

available on the display screen; otherwise, the 24h test shall be done again after the equipment has been repaired or replaced;

  2) The test process, repairing measures and test results shall be recorded.

10.4.7 The debugging and test of the call-intercom system shall meet the following requirements:

 1 The software system parameters, processing function and communication function of the allocated server, computer and master intercom, shall meet the design requirements;

 2 Debug each equipment, they shall achieve a correct service state;

 3 Each terminal of the system shall be encoded and its position shall be recorded in the software system;

 4 Debug the response state of the master intercom and paging terminal one by one in two-direction, ,it shall response correctly, and the glisten of the signal lamp shall be clear and distinct;

 5 Debug and test the display function of the system, and the displayed information of each display screen shall be accurate and clear;

 6 Debug and test the image and voice of the system terminal, the distortion shall be able to meet the design requirements;

 7 Debug and test the opening function of the system access guard, the access guard shall be able to correctly response to the opening order;

 8 During debugging and test, if the application software system has any error, the software shall be inspected and modified, and allocated and debugged again;

 9 All equipment shall be subjected to 24h uninterrupted functional and performance tests after system debugging, and shall meet the following requirements:

  1) During the test, if systematic fault or reliability fault occurring, 24h test shall be done again after equipment has been repaired or replaced;

  2) The test process, repairing measures and test results shall be recorded.

10.4.8 The debugging and test of the ticket checking and selling system shall meet the following requirements:

 1 The software system parameters, processing function and communication function of the allocated server, supervisory control computer, ticket-selling machine and card reading ticket-checking machine shall meet the design requirements;

 2 Debug and detect each function of the software, and it shall meet the design requirements;

 3 Debug the sensitivity of the card reading ticket-checking machine, and it shall accurately identify the information of the card and ticket, and write back correctly;

 4 The ticket checking system shall be able to accurately record the card reading and account charging information on the card reading ticket-checking machine;

 5 During debugging and test, if the application software system has any error, the software shall be inspected and modified, and then be allocated and debugged again;

 6 All equipment shall be subjected to 24h uninterrupted functional and performance tests after system debugging, and shall meet the following requirements:

  1) During the test, if systematic fault or reliability fault occurring, 24h test shall be done again after equipment has been repaired or replaced;

  2) The test process, repairing measures and test results shall be recorded;

7 The distinguish and treatment functions of the card reader for different kinds of cards, such as on/off gate card, cue card, record card, print card, etc., shall be tested.

10.4.9 Each system shall be subjected to commissioning after finishing debugging and test, and records for commissioning conditions and relevant data of system equipment inspection, installation and debugging shall be recorded.

10.5 Self-examination and Test

10.5.1 The inspection of each system shall meet the following requirements:

1 Each system shall be detected, and detection records and reports shall be filled in and prepared;

2 Documents for configuration parameters and instructions of the equipment and software shall be completed.

10.5.2 The inspection of the telephone switching system shall meet the following requirements:

1 The switching function of the system shall meet the requirements of normal conversation;

2 The maintenance and management functions of the system shall meet the requirements of the function provided by the system, and shall be detectable, manageable and repairable;

3 The signal aspect and network management functions of the system shall ensure correct signaling and the network management function shall meet the design requirements;

4 Inspection for the telephone switching system shall be carried out in accordance with those specified in Table 10.5.2.

Table 10.5.2 Inspection Items of the Telephone Switching System

Inspection before power on test		The nominal working voltage is -48V	The allowable variation scope is $-57V \sim -40V$
Hardware inspection and test		Visible and audible alarm single shall work normally	In accordance with the relevant regulations of the current professional standard YD 5077 *Accepting Specifications for SPC Exchange Installation Engineering*
		Load the test procedure, confirm that the hardware system is without failure by self-examination	
System inspection and test		Various calls, maintenance management, signal aspect and network support functions of the system	
Preliminary inspection and test	Reliability	Shall not result in more than 50% of the subscriber lines and junction lines failing in call	In accordance with the relevant regulations of the current professional standard YD 5077 *Accepting Specifications for SPC Exchange Installation Engineering*
		Call drops or stop connection of each user group shall not be large than 0.1 time / every month	
		Call drops or stop connection of the relaying group: 0.15 times/month (less than or equal to 64 speech path); 0.1 times/month (64 speech path $\sim$ 480 speech path)	
		Abnormal incoming call and outgoing call connection of the individual users: per thousand users shall be less than or equal to 0.5 user • times/month; per hundred relaying shall be less than or equal to 0.5 line • times/month	
		The restart index of the processor within one month shall be $1 \sim 5$ times (including three kinds of restart)	
		Software test fault shall not greater than 8 pcs/month, hardware as printed circuit board replacing times shall not greater than 0.05 times/100 user and 0.005 times/30 PCM systems per month	
		For a long time conversation, 12 intercoms shall maintain 48h	

Table 10.5.2(continued)

Inspection before power on test			The nominal working voltage is -48V	The allowable variation scope is $-57V \sim -40V$
Preliminary inspection and test	\multicolumn{2}{l	}{Obstacle rate test: the obstacle rate in the office shall not be greater than 3.4×10^{-4}}		40 users conduct simulated call for 100000 times simultaneously
	Performance test		Home exchange call	3 times to 5 times for inspecting measurement each time
			Outgoing and incoming exchange call	100% relaying test
			Tandem trunking test (various modes)	5 times for inspecting measurement of each mode
			Other types of exchange call	—
			The index of charging error rate shall not exceed 10^{-4}	—
			Special service business (extremely for 110, 119 and 120 etc.)	100% test
			The subscriber line is accessed into the modem and the transmission rate is 2400 bps, and the data error rate shall not be greater than 1×10^{-5}	—
			2B+D user test	—
	\multicolumn{3}{l	}{Relaying test: the repeat circuit call test, take $2 \sim 3$ circuits for inspecting measurement (including various ringing conditions)}	Mainly for signaling and interfaces
	Call completing rate test		The call completing rate between offices shall up to more than 99.96%	60 pairs of users, 100000 times
			The call completing rate between offices shall up to more than 98%	Call 200 times
	\multicolumn{3}{l	}{Adopt man-machine command to carry out with failure diagnosis test}	—

10.5.3 The inspection of the accessed network system shall meet the following requirements:

1 The transmissibility, signal aspect, physical interface and interface protocol of the communication system when accessing into the public network channel shall meet the design requirements;

2 The incoming and outgoing call of the exterior line shall be normal;

3 The inspection of the accessed network system shall meet the requirements of the contents specified in Table 10.5.3, and the inspection results shall meet the design requirements.

Table 10.5.3 Inspection Items of the Accessed Network System

Installation environment inspection		Machine room environment
		Power source
		Grounding resistance value
Equipment installation inspection		Pipeline laying
		Equipment cabinet and module
System inspection	Line interfaces of the transceiver	Power spectral density
		Longitudinal balance loss
		Overvoltage protection
	User network interface	25.6 Mbit/s electrical interface
		10 BASE-T interface
		USB interface
		PCI interface
	Service node interface	STM-1 (155 Mbit/s) optical interface
		Telecommunication interface
	Separator test	
	Transmission performance test	
	Functional verification test	Transmission function
		Management function

10.5.4 The inspection of the clock system shall meet the following requirements:

1 The system shall be possessed of monitoring functions to monitor the operation conditions of the systematic master clock, slave clock time server, time service, etc.;

2 The system shall be possessed of such control functions to control the synchronization between the master clock and timing signal receiver, and to control synchronous timing conducted by the master clock against the slave clock;

3 The system shall be able to automatically recover after the power-failure;

4 The clock system shall be possessed of the timing and time service function towards host machines of other intelligent systems;

5 Main technical parameters as the independent timing accuracy of master clock and the synchronous error of the master and slave clocks shall meet the design requirements.

10.5.5 The inspection of the information guidance and release system shall meet the following requirements:

1 The display accuracy and effectiveness of all menu items on the operation interface of the software of the local machine of the system shall be inspected one by one;

2 The networking function of the network play control, system configuration management and log information management of the system shall be inspected one by one;

3 The installation of system display equipment and the transmission line of the power supply shall be inspected.

10.5.6 The inspection of the call-intercom system shall meet the following requirements:

1 The response of the master intercom to each intercom terminal shall be in time and correct;

2 The audio effects of the call-intercom system shall be inspected;

3 The broadcasting and call performance of the call-intercom system shall be inspected by a sound pressure meter;

4 The image and voice of the call-intercom system shall be distinct;

5 The software management platform of the server and the workstation shall normally operate and the function shall be complete.

10.5.7 The inspection of ticket selling and checking system shall meet the following requirements:

1 The ticket vending machine and the card creator machine shall be able to finish ticket selling and card creating correctly, and the response time shall meet the design requirements;

2 The installation quality and reliability of the ticket checking access gate shall meet the design requirements; where a shear type baffle is used, the detection of the opening and closing strength shall meet the design requirements;

3 The ticket checking device shall be able to distinguish the ticket information accurately and reliably, and shall be able to response synchronously and write-back correctly;

4 The access gate shall be able to execute the switch commands of the system accurately and generate correspond electromechanical operations;

5 The center server of the ticket selling system shall classify the ticketing terminal data and record the summary statistics;

6 The software management platform of the server and the workstation shall normally operate and the function shall be complete.

10.6 Quality Record

10.6.1 Table B. 0. 12 in this code shall be filled in for the Quality Acceptance Record Sheet of the Telephone Switching System.

10.6.2 Table B. 0. 13 in this code shall be filled in for the Quality Acceptance Record Sheet of the Access Network Equipment.

10.6.3 Table B. 0. 14 in this code shall be filled in for the Quality Acceptance Record Sheet of the Clock System.

10.6.4 Table B. 0. 15 in this code shall be filled in for the Quality Acceptance Record Sheet of the Information Guidance and Release System.

10.6.5 Table B. 0. 16 in this code shall be filled in for the Quality Acceptance Record Sheet of the Call-Intercom System.

10.6.6 Table B. 0. 17 in this code shall be filled in for the Quality Acceptance Record Sheet of the Ticket Selling and Checking System.

11 Information Application System

11.1 General Requirements

11.1.1 This Chapter is applicable to implement, preparation, installation (installation of the software and hardware), debugging and self-examination and test of the office operation system, real property operation and management system, public service management system, public information service system, intelligent card application system, information network safety management system and application systems required by other service functions.

11.2 Construction Preparation

11.2.1 The technical preparation shall meet the following requirements:

1 According to the requirements of the design document, the construction organization shall finish the network planning, allocation plan, system function and performance files of the informational application system which shall be approved by the joint review;

2 Installation and commissioning manuals and technical parameter documents of the software and hardware products shall be available;

3 The construction organization shall finish the construction and commissioning plan of the system which shall be approved by the joint review.

11.2.2 The material and equipment preparation shall meet the following requirements:

1 The equipment and software shall be carried out with product quality inspection according to the requirements of Section 3.2 in the current national standard GB 50339 - 2003 *Code for Acceptance of Quality of Intelligent Building Systems*, and shall meet the requirements of the site acceptance;

2 The specifications, models, quantity and performance parameters of the server and workstation shall meet the requirements of the documents for system function and performance;

3 The quantity, visions and performance parameters of such basic software of the operating system, database and antivirus software shall meet the requirements of the documents for system function and performance;

4 The electronic documents or database of the basic service data of the users shall be collected.

11.2.3 The construction of the generic cabling system, information network system and other relevant information facility systems shall be completed.

11.3 Installation of the Hardware and Software

11.3.1 For software installation, software programming and development shall be carried out according to documents for system function and performance, and quality inspection shall be carried out for the application software according to the requirements of Section 6.2 in this code.

11.3.2 For software installation, system diagram, network topology diagram and equipment layout wiring diagram shall be made according to the network planning and allocation plan and the system function and performance document.

11.3.3 The installation of such equipment as server and workstation shall meet the requirements of Article 6.2.1 in this code.

11.3.4 Software irrelevant with this system shall not be installed and operated in the server and workstation.

11.3.5 The software debugging and modification shall be conducted on dedicated computer, and shall be subjected to version control.

11.3.6 Software at the service end of the system should be allocated as automatic operation mode when starting up.

11.3.7 The safety measures for software installation shall meet the following specifications:

    1 Antivirus software shall be installed on the server and workstation, and shall be set in operation state;

    2 The user passwords of the operating system, database and application software shall meet the following requirements:

        1) The length of the password shall not be less than 8 digits;

        2) The password should be the combination of capital and small letters, figures and punctuations;

    3 Identical user names and password combinations shall not be used among several servers and workstations or among several software;

    4 Operations, as virus and malware killing, shall be periodically carried out for the server and workstation.

11.4 Quality Control

11.4.1 Quality control for dominant item shall meet the following requirements:

    1 Patch programs of the latest version shall be installed for the operating system, database and antivirus software;

    2 During the process of starting, operating and closing, the software and equipment shall be free from run-time errors;

    3 After software has been modified, it shall pass the system test and regression test.

11.4.2 Quality control for general item shall meet the following requirements:

    1 The Internet address shall be allocated to such equipment as the server and workstation according to the network planning and allocation plan;

    2 The foundation-platform software and antivirus software of the operating system and database shall be possessed of formal software use (authorized) license;

    3 The operating system and antivirus software of the server and workstation shall be set in auto updated operation mode;

    4 The configuration parameters for such equipment as the server and workstation shall be recorded.

11.5 System Debugging

11.5.1 The debugging preparation shall meet the following requirements:

    1 The installation of equipment and software shall be finished and the parameter shall also be allocated;

 2 The service basic data or test data required by debugging shall be logged in.

11.5.2 In the process of system debugging, the uninterrupted operating software as required by design shall be always in the operation state.

11.5.3 The working condition and operating log of the software shall be inspected every day.

11.5.4 After normal operation of the software and equipment, they shall be subjected to function test.

11.5.5 After function test, they shall be subjected to performance test.

11.5.6 Where operation errors occurring in the process of debugging, or the system function or performance cannot meet the design requirements, the problem reporting sheet for the system debugging shall be filled in.

11.5.7 Before finishing system debugging, all problem reports shall be handled and the problem treatment records for the system shall be filled in.

11.5.8 The technical personnel of the user's organization shall get involved in the function test and performance test.

11.6 Self-examination and Test

11.6.1 The application software of the system shall be detected, and the test records and test reports shall be finished.

11.6.2 The system shall be subjected to network security detection, and the test records and test reports of the network security system shall be finished.

11.6.3 Documents for configuration plan and instructions of the equipment and software shall be complete.

11.6.4 After system test, all tested users and tested data shall be deleted.

11.7 Quality Record

11.7.1 Table B.0.18 in Appendix B of this code shall be filled for the Information Application System Function Sheet.

11.7.2 Table B.0.19 in Appendix B of this code shall be filled in for the Configuration Parameter Records for the Information Application System.

12 Building Automation System

12.1 Construction Preparation

12.1.1 In addition to the requirements of the current national standard GB 50339 *Code for Acceptance of Quality of Intelligent Building Systems* and Article 3.3.2 in this code, the material and equipment preparation shall also meet the following requirements:

1 The model and material of electric valve shall meet the design requirements; after sampling test, the valve body strength and the spool leakage shall meet the requirements of the product instructions;

2 The input voltage, output signal and wiring mode of the electric valve drive shall meet the requirements of the design and the product instructions;

3 The drive stroke, pressure and maximum closing force of electric valve shall meet the requirements of the design and the product instructions; if necessary, they should be inspected by a third-party inspection organization;

4 The measuring devices (instruments) for temperature, pressure, flow and electricity shall be calibrated according to the relevant requirements; if necessary they should be inspected by a third-party test organization.

12.1.2 In addition to the requirements of Article 3.3.4 in this code, the construction environment shall also meet the following requirements:

1 After the civil work and decoration of the control room, weak current room and relevant equipment machine room of the building automation system have been completed, the machine room shall be provided with reliable power supply and grounding terminal block;

2 The installation of air conditioning unit, fresh air handling unit, blower/exhaust fan, water chiller, cooling tower, heat exchanger, water pump, pipeline, valve etc. shall be completed;

3 The installation of power transformation/distribution equipment, high and low voltage distribution cabinet, power distribution box, lighting distribution box etc. shall be completed;

4 The installation of water supply, drainage, fire pump, pipeline, valve etc. shall be completed;

5 The installation of the elevator and escalator shall be completed.

12.2 Equipment Installation

12.2.1 The requirements of this section are applicable to the installation of the following building automation system equipment:

1 Control center equipment such as console, net control unit, server, workstation, etc.;

2 Various sensors for temperature, humidity, pressure, pressure difference, flow, air quality, etc.;

3 Actuators such as electric air valve, electric water valve, electricmagnetic valve, etc.;

4 Field control unit, etc..

12.2.2 The installation of control center equipment shall meet the following requirements:

1 The installation position of console shall meet the design requirements; the installation shall be stable, firm and convenient for operation and maintenance;

2 The rack, wiring and grounding in the console shall meet the design requirements;

3 The net control unit should be firmly installed on the rack in the console;

4 Equipment such as server, workstation and printer shall be installed according to the requirements of construction drawings and be arranged stably in order;

5 The connection of the power cable, communication cable and control cable of the control center equipment shall meet the design requirements; the wires and cables shall be arranged in order, avoided crossing and set with proper marking.

12.2.3 The installation of control center software shall meet the requirements of Article 6.3.2 in this code.

12.2.4 The installation of field control unit box shall meet the following requirements:

1 The installation position of field control unit cabinet should be close to the electric cabinet of the controlled equipment;

2 The field control unit cabinet shall be installed firmly and shall not tilt; where it is installed on the light-weight wall, reinforcement measures shall be taken;

3 Where the height of the field control unit cabinet is not larger than 1 m, it should be hung on the wall; the height from the cabinet center to the floor shall not be less than 1.4 m;

4 Where the height of the field control unit cabinet is greater than 1 m, it should be installed on the floor and set with a pedestal;

5 The clear distance from the side face of the field control unit cabinet to the wall or other equipment shall not be less than 0.8 m; and the front operating distance shall not be less than 1 m;

6 The field control unit cabinet shall be wired according to the wiring diagram and equipment instruction manual; wiring shall be fixed firmly in order and should not cross with each other; the ends shall be numbered;

7 Wiring diagram for the equipment in the cabinet shall be attached to the inner side of the door panel of field control unit cabinet;

8 The field control unit shall be installed, kept properly and provided with dust-proof, damp proof and anti-corrosive measures prior to debugging.

12.2.5 The installation of indoor and outdoor temperature/humidity sensors shall meet the following requirements:

1 The installation position of indoor temperature/humidity sensors should keep a distance greater than 2 m to the doors, windows and air outlets; as for the indoor temperature/humidity sensors installed in the same area, their distance to the floor shall be consistent and the height difference shall not be greater than 10 mm;

2 The outdoor temperature/humidity sensors shall be provided with wind-proof and rain-proof measures;

3 The indoor and outdoor temperature/humidity sensors shall not be installed at the places directly exposed to the sunlight and shall be kept away from the area with strong vibration, electromagnetic interference and damp.

12.2.6 The duct type temperature/humidity sensors shall be installed at the lower part of straight pipe section with stable wind speed.

12.2.7 The installation of water pipe temperature sensor shall meet the following requirements:

1 It shall be installed orthogonal to the pipeline and its axis shall be vertically crossed with the pipeline axis;

2 Where the temperature sensing section is less than 1/2 of the pipeline caliber, it shall be installed at the side face or bottom of the pipeline.

12.2.8 The duct type pressure sensor shall be installed at the upper part of the pipeline and at the upstream pipe section of the temperature measuring point of temperature/humidity sensor.

12.2.9 The water pipe type pressure/differential pressure sensor shall be installed at the upstream pipe section of the pipeline of temperature sensor; where the pressure measuring section is less than 2/3 of the pipeline caliber, it shall be installed at the side or bottom of the pipeline.

12.2.10 The installation of differential pressure switch for wind pressure shall meet the following requirements:

1 Airtight treatment shall be carried out after installation;

2 The installation height should not be less than 0.5m.

12.2.11 The water flow switch shall be installed vertically on the horizontal pipeline section. The direction of arrow indicated on the water flow switch shall be consistent with the direction of water flow. The length of water flow vane shall be greater than 1/2 of the pipe diameter.

12.2.12 The installation of water flow sensor shall meet the following requirements:

1 The distance from the installation position of the water flow sensor to the valve, pipeline necking and the bent pipe shall not be less than 10 times of the pipeline internal diameter;

2 The water flow sensor shall be installed at the upstream of the pressure measuring point with a distance of 3.5~5.5 times of the pipe inside diameter to the pressure measuring point;

3 The water flow sensor shall be installed at the upstream of the temperature measuring point of the temperature sensor with a distance of 6~8 times of the pipe diameter to the temperature sensor;

4 The transmission line of flow sensor signal should be adopted shielded or insulation sheathed cables; the shielding layer of the wire cable should be grounded at the field control unit side.

12.2.13 The installation of the indoor air quality sensor shall meet the following requirements:

1 The air quality sensor detecting light gas specific gravity shall be installed at the upper part of the room and the installation height should not be less than 1.8m;

2 The air quality sensor detecting heavy gas specific gravity shall be installed at the lower part of the room and the installation height should not be greater than 1.2m.

12.2.14 The installation of the duct type air quality sensor shall meet the following requirements:

1 The duct type air quality sensor shall be installed at the horizontal straight pipe section of the duct pipeline;

2 The air quality sensor detecting light gas specific gravity shall be installed at the upper part of the duct;

3 The air quality sensor detecting heavy gas specific gravity shall be installed at the lower part of the duct.

12.2.15 The installation of the air valve actuator shall meet the following requirements:

1 The connection between the air valve actuator and the air valve axis shall be fixed firmly;

2 The mechanical mechanism of the air valve shall open and close flexibly and be free from looseness or jamming phenomenon;

3 Where the air valve actuator cannot be directly connected with the air door baffle shaft, it may be connected with the baffle shaft through accessories. However, the accessory device shall ensure the adjustment range of the rotation angle of air valve actuator;

4 The output torque of the air valve actuator shall match with the torque required by the air valve and shall meet the design requirements;

5 The open/close indicating bit of the air valve actuator shall be consistent with the actual state of the air valve and the air valve actuator should face to the position convenient for observation.

12.2.16 The installation of electric water valve and electricmagnetic valve shall meet the following requirements:

1 The direction of arrow on the valve body shall be consistent with the direction of water flow; the valve shall be installed vertically on the horizontal pipe;

2 The valve actuator shall be installed firmly with flexible transmission but no looseness or jamming phenomenon; the valve shall be located at the position convenient for operation;

3 As for the valves with valve position indicating device, the valve position indicating device shall face the position convenient for observation.

12.3 Quality Control

12.3.1 The dominant items shall meet the following requirements:

1 If welding is required in the sensor installation, it shall meet the relevant requirements of the current national standard GB 50236 *Code for Construction and Acceptance of Field Equipment, Industrial Pipe Welding Engineering*;

2 The leading inlet of the sensor and actuator junction box should not be upward; where it is inevitable, sealing measures shall be taken;

3 The installation of the sensor and actuator shall be carried out strictly according to the requirements of the instructions; the wiring shall be carried out in accordance with the wiring diagram and the equipment instruction manual; the wirings shall be fixed firmly in order and should not cross with each other; their ends shall be numbered;

4 The water pipe type temperature sensor, water pipe pressure sensor, water flow switch and water pipe flowmeter shall be installed at the straight pipe section with stable water flow, shall avoid dead corner of water flow stream and should not be installed at the pipeline weld seam;

5 The duct type temperature/humidity sensor, pressure sensor and air quality sensor shall be installed at the straight pipe section of the duct where the gas flow stream is stable and shall avoid the ventilation dead corner in the duct;

6 The shielding layer of the instrument cable and wire shall be grounded at the instrument panel side in the control room; the shielding layer of the same circuit shall be possessed of reliable electrical continuity and shall not be suspended in air or be grounded repeatedly.

12.3.2 The general items shall meet the following requirements:

1 The installation quality of field equipment (e. g. sensor, actuator and control cabinet) shall meet the design requirements;

2 The number of wirings of each terminal in the control unit cabinet terminal board shall not exceed two;

3 Neither the sensor nor the actuator shall be covered by the thermal insulation materials;

4 The sensors and differential pressure switches for duct pressure, temperature, humidity, air quality, air speed etc. shall be installed after the thermal insulation and purging of duct are completed;

5 The sensor and actuator should be installed at the well-lighted position where is convenient to operate; the sensor and actuator shall be avoided to be installed at the position with vibration, damp, prone to mechanical damage, strong electromagnetic field interference and high temperature;

6 Knock or vibration shall not be in the installation process of the sensor and actuator; they shall be installed firmly and flatly; the various members used for installing the sensor and actuator shall be connected firmly, under uniform stress and treated for rust prevention;

7 The installation of the water pipe type temperature sensor, water pipe type pressure sensor, steam pressure sensor and water flow switch should be carried out simultaneously with the installation of the processing pipeline;

8 Tapping and welding of the installation sleeves for water pipe type pressure, pressure differential and steam pressure sensors as well as water flow switch and water pipe flowmeter shall be carried out before the corrosion protection, lining, purging and pressure test of the processing pipeline;

9 Where the fan coil temperature control unit is installed in parallel with other switches, the height difference shall be less than 1mm; if they are installed in the same room, the height difference shall be less than 5mm;

10 The valves and actuators installed outdoors shall be provided with sun-proof and rain-proof measures;

11 The normally uncharged metal parts such as the enclosure, instrument box, cable trough, support and pedestal of the power using instrument shall be provided with protective grounding;

12 The signal circuit grounding and shield grounding of the instrument and control system shall be share one grounding wire.

12.4 System Debugging

12.4.1 The debugging preparation shall meet the following requirements:

1 The installation of the control center equipment and software shall be completed; the cable laying and wiring shall meet the requirements of the design and the product instructions;

2 The installation of the field control unit shall be completed; the cable laying and wiring shall meet the requirements of the design and the product instructions;

3 The installation of various actuators and sensors shall be completed; the cable laying and wiring shall meet the requirements of the design and the product instructions;

4 The communication interface and cable laying between the building automation system equipment and subsystem (equipment) shall meet the design requirements;

5 The controlled equipment and its system shall be installed, debugged and able to be operated normally;

6 The power supply and grounding of the building automation system equipment shall meet the design requirements;

7 The net control unit shall normally communicate with the server and the workstation. The power supply of the net control unit shall be connected to the uninterruptible power supply to ensure the regular power supply for net control unit during the debugging period;

8 The field control unit program shall be written and shall meet the design requirements.

12.4.2 The debugging of the field control unit shall meet the following requirements:

1 The resistance between the ground pin and all I/O port terminals shall be measured and be greater than 10kΩ;

2 No AC voltage between the ground pin and all I/O port terminals shall be confirmed;

3 The debugging instrument and the field control unit shall be able to communicate normally; the parameters of other field control units shall be able to be checked through the busbar;

4 All digital input points shall be tested manually and be recorded;

5 All digital output points shall be tested manually and be recorded; the controlled equipment shall operate normally;

6 The type of analog input and output, the measuring range and the set value shall be determined and shall meet the requirements of the design and the equipment instruction manual;

7 According to the requirements of different signals, all analog inputs shall be tested manually and the test values be recorded;

8 All analog outputs shall be tested manually; the controlled equipment shall operate normally and the test values shall be recorded.

12.4.3 The group control debugging of the cold/heat source system shall meet the following requirements:

1 In automatic control mode, the start, stop and automatic exit sequence of the system equipment shall meet the design and processing requirements;

2 The cold/heat source system be able to automatically control the quantity of the cold and heat engine groups put into operation according to the change of cold and heat load;

3 Simulating the failure of a set of unit or water pump, the system shall be able to automatically start the standby unit or water pump and put it into operation;

4 The cold/heat source system be able to automatically control the quantity of cooling tower fans put into operation and control the open/close of the relevant electric water valves according to the temperature change of the return cooling water;

5 The cold/heat source system shall be able to automatically control the bypass valve according to the pressure difference change of the supply and return water;

6 The operation state of the water pump shall be able to be judged according to the indication of water flow switch state;

7 The cold/heat source system shall be able to automatically accumulate the start times and operation time of the equipment and be able to automatically prompt equipment overhaul at regular intervals;

8 The building automation system shall communicate normally to the water chilling unit control device; the various parameters of the water chilling unit shall be able to be collected normally.

12.4.4 The debugging of the air conditioning unit shall meet the following requirements:

1 Analog input values of temperature, humidity and wind pressure shall be inspected and be accurate. The state of such digital inputs as wind pressure switch and anti-freezing switch shall be normal and recorded;

2 Where the digital output parameters are changed, the open and close actuation of such equipment as fan, electric air valve, electric water valve, etc. shall be normal. Where the analog output parameters are changed, the actuation of the relevant air valve and electric control valve shall be normal; meanwhile, the position control shall change correspondingly and be recorded;

3 Where the filter pressure difference exceeds the set value, the differential pressure switch shall be able to give an alarm;

4 Simulating the alarm signal of the anti-freezing switch, the fan and the fresh air valve shall be able to automatically close and be recorded;

5 The air conditioning unit shall be able to automatically control the opening degree of the fresh air valve according to the change of carbon dioxide concentration;

6 The fresh air valve, fan and water valve shall be capable of automatic interlocking control;

7 Where the humidity set value is changed manually, the system shall be able to automatically control the open/close of the humidifier;

8 The system shall be able to automatically control the control program according to the reversal of seasons.

12.4.5 The debugging of the fan coil shall meet the following requirements:

1 Where the temperature set value and the mode setting of the temperature control unit are changed, the fan and the electric water valve shall operate normally;

2 Where the fan coil control unit and the field control unit are jointly debugged, the field control unit shall be able to change the temperature set value, control the start/stop of the fan and monitor the operation parameters.

12.4.6 The debugging of the blower/exhaust fan shall meet the following requirements:

1 The unit shall be able to automatically control the start/stop of the fan according to the control schedule;

2 The unit shall be able to automatically start/stop the fan according to the carbon monoxide and carbon dioxide concentration as well as the air quality;

3 Where the smoke exhaust fan is controlled jointly by the fire-fighting system and building automation system, the priority mode of fire-fighting control shall be realized.

12.4.7 The debugging of the water supply and drainage system shall meet the following requirements:

1 The parameters such as liquid level and pressure shall be inspected; the monitoring and alarming of water pump operation state shall be tested and recorded;

2 The system shall be able to automatically start/stop the water pump according to the water level in the water tank.

12.4.8 The debugging of the power transformation and distribution system shall meet the following requirements:

 1 The data read by the workstation and the data measured at the site shall be inspected; the graphic display function of parameters such as voltage, current, active (reactive) power, power factor and quantity of electricity shall be verified;

 2 The data read by the workstation shall be inspected and the alarm signals of transformer, generating unit, distribution box and distribution cabinet shall be verified.

12.4.9 The debugging of the lighting system shall meet the following requirements:

 1 The lighting circuits are controlled through the workstation; the switch and state of each lighting circuit shall be normal and shall meet the design requirements;

 2 The switch automatically controlling the lighting circuit according to the time schedule and the indoor and outdoor illuminance shall meet the design requirements.

12.4.10 According to the design requirements, the workstation shall verify the graphic display function of each operation parameter of the elevator.

12.4.11 The system joint debugging shall meet the following requirements:

 1 The model & specification of the connecting and transmission lines between such equipment as the control center server, workstation, printer, net control unit and communication interface (including other subsystems) shall be inspected and be correct without error;

 2 The communication protocol, data transmission format and rate of the communication interface shall meet the design requirements; the communication interface shall be able to communicate normally;

 3 The building automation system server, workstation management software and database shall be configured normally and the software function shall meet the design requirements;

 4 The monitoring performance and linkage function of the building automation system shall meet the design requirements.

12.5 Self-examination and Test

12.5.1 The inspection of server and workstation shall meet the following requirements:

 1 The installation of the server, workstation, net control unit and auxiliary equipment shall be inspected and shall meet the requirements of the design drawings;

 2 The change of each parameter in site shall be observed at the workstation and the status data shall be updated continuously;

 3 The analog output or digital output are controlled through the workstation, the site actuator or controlled object shall actuate correctly and effectively;

 4 When simulating the failure at the input side of the site control unit, the workstation shall log in the alarm failure data and give out sound for warning;

 5 When simulating power loss of the server or workstation, the server or workstation shall be able to automatically restore full monitoring and management function after the power is restored;

 6 The server setting software shall assign operation permissions and roles for the operation personnel;

 7 The software functions shall be complete, the human machine interface shall be in Chinese and the operation shall be easy and visualized;

8 The server shall be able to print the information for equipment operation time, area, number and state in the form of statement, diagram and tendency chart.

12.5.2 The inspection of field control unit shall meet the following requirements:

1 The installation of the field control unit cabinet shall be standardized, reasonable and easy to maintain;

2 Simulating the stop state of server and workstation, the field control unit shall be able to work normally;

3 Changing the set value of the controlled equipment, the actuation order/trend of the corresponding actuator shall meet the design requirements;

4 Simulating the power loss of the field control unit, the control unit shall be able to automatically restore the set operation state before power loss after the power is restored;

5 Simulating the interruption of the communication network of the field control unit and server, the field equipment shall be able to maintain the normal automatic operation state and the workstation shall be provided with fault alarm signal for control unit offline;

6 When the controlled equipment is started or stopped, the actuation order of the relevant equipment and actuator shall meet the design requirements;

7 The field control unit clock shall be synchronized with the server clock.

12.5.3 The inspection of sensor and actuator shall meet the following requirements:

1 The installation of the sensor and actuator on site shall be inspected and be standardized, reasonable and easy to maintain;

2 The data and state displayed at the workstation shall be inspected and be consistent with the reading and state on site;

3 The actuation or actuation sequence of the actuator shall be inspected and shall conform to the design process;

4 The zero opening state of the regulating valve shall be inspected;

5 When the parameter exceeds the allowable range, alarm signals shall be generated;

6 The control actuator at the workstation shall be able to actuate normally.

12.5.4 The group control inspection of cold/heat source system shall meet the following requirements:

1 The cold/heat source system shall be able to realize load regulation, preset time schedule based automatic start/stop and optimized control of energy conservation;

2 Where the time program is changed or the cold/heat source system is started or stopped manually through the workstation, the unit shall be able to operate normally according to the linkage control sequence;

3 The inspection system shall be able to stabilize the pressure difference between the water collector and header within the allowable range of the design by regulating the bypass valve;

4 The workstation shall be able to display the operation parameters of the cold/heat source system equipment and be able to record them automatically.

12.5.5 The inspection of air conditioning and ventilation system shall meet the following requirements:

1 The measured values of temperature and humidity displayed at the workstation shall be consistent with the on-site measured values of the portable temperature and humidity sensor;

2 The states of wind pressure difference switch and anti-freezing switch shall be inspected; and the accuracy of the alarm signal shall be checked by changing the set value manually;

3 The working state, control stability, response time and control effect of the fan, water valve and air valve shall be inspected;

4 Where the schedule is changed at the workstation, the detection system shall be possessed of automatic start/stop function;

5 The set values of the temperature and humidity shall be changed at the workstation, the temperature control process shall be recorded; the correctness, system stability, system response time and control effect of the linkage control program as well as the historical record of the system operation shall be inspected;

6 The failures shall be simulated, including the filter differential pressure switch alarm, fan failure alarm and temperature sensor over-limit alarm, and the accuracy and response time of the alarm signal shall be inspected at the workstation;

7 The operation states of blower/exhaust fan shall be monitored and the start and stop of blower/exhaust fan shall be able to be automatically controlled according to air parameters;

8 Fire-fighting linkage test shall be carried out for the air conditioning and ventilation system; where the fire alarm system gives an alarm, the operation of air conditioning and ventilation system shall meet the requirements of relevant specifications and design.

12.5.6 The inspection of water supply and drainage system shall meet the following requirements:

1 The start/stop control, operation state, fault alarm and liquid level of the water supply and drainage equipment shall be able to be controlled and monitored remotely through the workstation and be recorded;

2 Simulating the water level raising or lowering, the liquid level switch shall actuate normally and shall be able to be linked to the start or stop of water pump according to the control processing.

12.5.7 The inspection of power transformation and distribution system shall meet the following requirements:

1 The measured values of such parameters as the voltage, current, active (reactive) power, power factor and quantity of electricity of the power transformation and distribution system shall be compared with the workstation readings for the inspection of accuracy and authenticity;

2 The working states and failures of high- and low-voltage switch cabinets, transformer and generating unit shall be monitored;

3 The dynamic graphics of each parameter at the workstation shall be able to reflect the parameter change accurately.

12.5.8 The inspection of public lighting system shall meet the following requirements:

1 The lighting equipment shall be monitored and the accuracy of control actuation shall be inspected on basis of the outdoor illuminance and time schedule;

2 The control function on all lighting circuits through workstation shall be inspected.

12.5.9 The inspection of elevator and escalator systems shall meet the following requirements:

1 The workstation shall be set with elevator's dynamic simulation chart to display the current position, operation state and failure alarm of the elevator;

2 The operation parameter of the elevator system monitored by the workstation shall be inspected and also be verified with the actual state.

12.5.10 The inspection of system real-time performance and reliability shall meet the following requirements:

1 The alarm signal response time, detection system sampling speed and response time shall be recorded by such detection instruments as stopwatch and also shall meet the design requirements;

2 Where one or more field control units lose power, the workstation shall be able to output correct alarm;

3 Simulating the power failure of the server and workstation, the communication busbar and site control unit shall be able to work normally and will not influence the normal operation of the controlled equipment.

12.6 Quality Record

12.6.1 Table B.0.20 in this code shall be filled in for the Test Record of Control Unit Wire Cable.

12.6.2 Table B.0.21 in this code shall be filled for the Single Point Debugging Record.

13 Fire Alarm and Control System

13.1 Construction Preparation

13.1.1 The construction of the fire alarm and control system must be undertaken by the construction organization with corresponding qualification grade.

13.1.2 Where the fire alarm and control system is integrated with the emergency command system and intelligentized integration system, it shall provide communication interface and protocol to external users and shall meet the requirements of Article 15.1.1 in this code.

13.1.3 The material and equipment preparation shall meet the following requirements:

1 The main equipment and materials of the fire alarm and control system shall be selected according to the design requirements and meet the requirements of Section 2.2 in GB 50166 - 2007 *Code for Installation and Commissioning of Automatic Fire Alarm System*;

2 Where the fire emergency broadcast and public address system share the same set of system, the shared equipment of the public address system shall be product passing the national certification (approval) and the product name, model and specification shall be consistent with the inspection report;

3 The bridge, wire cable, steel pipe, metal hose, flame retardant plastic pipe, fire-retardant coating and installation accessories, etc. shall meet the requirements of fire protection design;

4 The variety and voltage grade of the wire cable shall be inspected according to the relevant requirements of the current national standard GB 50116 *Code for Design of Automatic Fire Alarm System*.

13.2 Equipment Installation

13.2.1 In addition to the requirements of Section 3.2 in GB 50166 - 2007*Code for Installation and Acceptance of Fire Alarm System* and the requirements of Chapter 4 in this code, the bridge and pipeline laying shall also meet the following requirements:

1 The wire cable of fire alarm and control system shall be laid by means of bridge and dedicated wire pipe;

2 Connection of alarm wire cables shall be done in the terminal box or dividing box, and conductor connection shall be adopted reliable crimping or welding;

3 Protective grounding shall be carried out for the bridge and metal wire pipe.

13.2.2 In addition to the requirements in Sections 3.3~3.10 in the current national standard GB 50166 - 2007 *Code for Installation and Acceptance of Fire Alarm System*, the equipment installation shall also meet the following requirements:

1 The terminal box and module box should be arranged in the weak-current room and be fixed on the wall according to the design height; the terminal box and module box shall be installed upright and firmly;

2 The trunk line led out from the fire control room and the control circuit of the fire alarm and other equipment shall be bundled respectively and collected on both sides of the terminal board; on the left side is the trunk line and on the right side is the control circuit.

13.2.3 In addition to the relevant requirements of GB 50166 *Code for Installation and Acceptance of Fire Alarm System*, the equipment grounding shall also meet the following requirements:

   1 The working grounding wire shall be adopted copper-cored insulated wire or cable and shall not use galvanized strap iron or metal hose;

   2 The enclosure and base of the fire control equipment shall be grounded reliably; the grounding wire shall be led to the grounding terminal box;

   3 The fire control room shall be set with dedicated grounding box for working grounding according to the design requirements. The ground resistance shall meet the requirements of 16.2.1 in this code;

   4 The protective grounding wire and working grounding wire shall be separated and the metal hose shall not be used as the protective grounding conductor.

13.3 Quality Control

13.3.1 The dominant items shall meet the following requirements:

   1 The category, model, position, quantity, function and the like of the detectors, modules and alarm buttons shall meet the design requirements;

   2 The model, position, quantity, function and the like of the fire telephone jacks shall meet the design requirements;

   3 The position, quantity, function and the like of fire emergency broadcast shall meet the design requirements and it shall be able to cut off the public broadcasting system and broadcast the fire accident within 10s after the manual or alarm signal is triggered;

   4 The function and model of the fire alarm control unit shall meet the design requirements;

   5 The interlocking between the fire alarm and control system and the firefighting equipment shall meet the design requirements.

13.3.2 The general items shall meet the following requirements:

   1 The detector, module and alarm button shall be installed firmly, well equipped and be free from damage and deformation;

   2 The wires connection of the detector, module and alarm button shall be crimped or welded reliably and be provided with marks; allowance shall be made for the external wires;

   3 The installation position of the detector shall meet the requirements of the protection radius and area.

13.4 System Debugging

13.4.1 System debugging shall be implemented according to the requirements of Chapter 4 in the current national standard GB 50166-2007 *Code for Installation and Acceptance of Fire Alarm System*.

13.5 Self-examination and Test

13.5.1 Preparation for the system self-examination and test shall meet the following requirements:

1 The system self-examination and test shall be carried out after the system installation debugging;

2 The system equipment and circuit wiring shall be correct; the insulation of all circuits and electrical equipment shall be inspected and be free from looseness, cold solder joint, wrong wire or shedding phenomena; any discovered problem shall be treated and recorded;

3 The system self-examination and test shall be coordinated with the relevant profession and the linkage equipment of the relevant profession shall be in the normal working condition.

13.5.2 The system self-examination and test shall meet the following requirements:

1 Firstly, single machine power-on inspection shall be carried out for the device and equipment one by one (including alarm controller, linkage control panel, fire broadcasting, etc.); when the device and equipment are normal, system inspection may be carried out;

2 When the fire alarm and control system is powered on, the equipment function shall be tested according to the requirements of the current national standard GB 16806 *Automatic Control System for Fire Protection*;

3 After the single machine inspection and each fire-fighting equipment inspection, system linkage inspection shall be carried out;

4 Where the public broadcasting system is shared by the fire emergency broadcasting, the broadcasting system shall be able to be switched and broadcast the fire accident within 10s after the manual or alarm signal is triggered;

5 The linkage between the fire alarm and control system and the security system shall meet the requirements of Article 13.4.7 in the current professional standard JGJ 16-2008 *Code for Electrical Design of Civil Buildings*.

13.6 Quality Record

13.6.1 In addition to the requirements of Section 3.7 in this code, the quality record of fire alarm and control system shall also comply with the relevant requirements the current national standard GB 50166 *Code for Installation and Acceptance of Fire Alarm System*.

14 Security System

14.1 Construction Preparation

14.1.1 Equipment such as matrix switching control unit, digital matrix, network switch, video camera, control unit, alarm detector, storage device and display device shall be provided with compulsory product certification and "CCC" sign or documents such as network access license, quality certificate and test report. The product name, model and specification shall be consistent with the inspection report.

14.1.2 The imported equipment shall be provided with relevant inspection certificate issued by the national commodity inspection department. All accompanied source materials and the design calculation materials, drawings, test records and acceptance evaluation conclusion of home-made equipment shall be counted, sorted and filed.

14.2 Equipment Installation

14.2.1 Laying of metal wire slot, steel pipe and wire cable shall meet the requirements of Chapter 4 in this code and Section 3.3 of the current national standard GB 50198 - 94 *Technical Code for Project of Civil Closed Circuit Monitoring TV System*.

14.2.2 The installation of video surveillance & control system shall meet the following requirements:

    1 The equipment installation and wire cable laying in the surveillance & control center shall comply with Section 3.4 in the current national standard GB 50198 - 94 *Technical Code for Project of Civil Closed Circuit Monitoring TV System*;

    2 The laying spacing of strong- and weak-current cables in the surveillance & control center shall meet the requirements of Article 2.3.8 in the current national standard GB 50198 - 94 *Technical Code for Project of Civil Closed Circuit Monitoring TV System* and shall be provided with obvious permanent signs;

    3 In addition to those specified in Article 6.3.5 in the current national standard GB 50348 - 2004 *Technical Code for Engineering of Security and Protection System*, Section 3.2 in GB 50198 - 94 *Technical Code for Project of Civil Closed Circuit Monitoring TV System* and Article 14.3.3 in JGJ 16 - 2008 *Code for Electrical Design of Civil Buildings*, the installation of camera, PTZ and decoder shall also meet the following requirements:

        1) Before installation, the video camera and lens shall be powered on for inspection and should operate normally;

        2) The equipment safety shall be considered when determining the installation position of video camera, and its view field shall not be blocked;

        3) When the overhead line is led into the PTZ, the bending of drip elbow shall not be less than the minimum bending radius of electrical (optical) cable;

        4) Rain-proof, anti-corrosive and lightning protection measures shall be taken when the outdoor video camera and decoder are installed;

4 The installation of optical transmitter, encoder and equipment box shall meet the following requirements:

    1) The optical transmitter and decoder shall be installed in the equipment box near the video camera; the equipment box shall be dustproof, waterproof and anti-theft;

    2) Before installation, the video encoder shall be connected with the front video camera for test; and the video encoder may be installed when the image transmission and data communication are normal;

    3) In the equipment box, the equipment shall be arranged in order and the wire route shall be provided with signs and wiring diagram;

5 Installation of application software shall meet the requirements of Article 6.2.2 in this code.

14.2.3 In addition to the requirements of Article 6.3.5 in the current standard of the nation GB 50348-2004 *Technical Code for Engineering of Security and Protection System* and Section 14.2 in JGJ 16-2008 *Code for Electrical Design of Civil Buildings*, the installation of intrusion alarm system equipment shall also meet the following requirements:

1 The detector shall be installed firmly and no obstacle shall exist in the detection range;

2 The outdoor detector shall be installed at the positions of dry, ventilated, free of accumulated water and provided with waterproof and moisture-proof measures;

3 Magnetically controlled switch should be installed on the door or window; the installation shall be firm, in order and beautiful;

4 The installation position of vibration detector shall be far away from such vibration sources as motor, water pump, water tank, etc.;

5 The installation position of glass break detector shall be close to the protection target;

6 The installation position of emergency button shall be concealed, convenient for operation and firm;

7 In the installation of infrared detector, the receiving end shall be installed along the light direction and avoid direct sun light and other high-power direct lights.

14.2.4 In addition to the relevant requirements of the current national standard GB 50396 *Code of Design for Access Control Systems Engineering*, the installation of access control system equipment shall also meet the following requirements:

1 The installation position of reader device shall avoid such severe environment as strong electromagnetic radiation source, damp and corrosivity;

2 The control unit and card reader shall not share the power socket with large current equipment;

3 The control unit should be installed in the weak-current room etc. places where are convenient for maintenance;

4 After the installation of card reader type equipment, protective structure surface shall be provided and be able to defense destructive attacks and technical opening;

5 The distance between the control unit and card reader should not be greater than 50m;

6 The supporting lockset shall be installed firmly and its open/close shall be flexible;

7 The infrared optoelectronic device shall be installed firmly; the sending and receiving devices shall be aligned with each other and shall avoid direct sun light;

8 In the installation of signal lamp control system, the distance between the warning light and detector shall not be greater than 15m;

9 The installation of access control system equipment adopting biometric identification technologies such as human face, eye pattern, fingerprints and palm prints to identify and read shall meet the requirements of product technical specification.

14.2.5 In addition to the requirements of Item 8 in Article 6.3.5 of the current standard of the nation GB 50348 - 2004 *Technical Code for Engineering of Security and Protection System* and Section 14.6 in JGJ 16 - 2008 *Code for Electrical Design of Civil Buildings*, the installation of parking garage (lot) management system shall also meet the following requirements:

1 The buried position of induction coil shall be centered and the center distance from card reader and gate machine should be 0.9~1.2m;

2 The vehicle stop shall be installed firmly and flat; where it is installed outdoors, waterproof, anti-collision and anti-smashing measures shall be taken;

3 The parking spot status signal indicator shall be installed at the visible position of the lane entrances/exits; the installation height shall be 2.0m~2.4m; where it is installed outdoors, waterproof and anti-collision measures shall be taken.

14.2.6 The installation of visitor (video) intercom system shall comply with the requirements of Item 6 in Article 6.3.5 of the current national standard GB 50348 - 2004 *Technical Code for Engineering of Security and Protection System*.

14.2.7 The installation of electronic guard tour management system shall comply with the requirements of Item 7 in Article 6.3.5 of the current national standard GB 50348 - 2004 *Technical Code for Engineering of Security and Protection System* and Section 14.5 in JGJ 16 - 2008 *Code for Electrical Design of Civil Buildings*.

14.2.8 The installation of security system control unit shall comply with the requirements of Item 9 in Article 6.3.5 of the current national standard GB 50348 - 2004 *Technical Code for Engineering of Security and Protection System*.

14.2.9 The construction of power supply, lightning protection and grounding system shall comply with the requirements of Article 6.3.6 in the current national standard GB 50348 - 2004 *Technical Code for Engineering of Security and Protection System* and Chapter 16 in this code.

14.3 Quality Control

14.3.1 The dominant items shall meet the following requirements:

1 The installation of the main equipment of each system shall be installed firmly and correctly wired; meanwhile effective anti-interference measures shall be taken;

2 The interconnection and interchange of the systems shall be inspected and the linkage between subsystems shall meet the design requirements;

3 The image quality and saving time recorded by the surveillance & control center system shall meet the design requirements;

4 Equipotential bonding shall be carried out for surveillance & control center. the grounding resistance shall meet the design requirements.

14.3.2 The general items shall meet the following requirements:

1 Terminating of each equipment and device shall be standardized;

2 The video image shall be free from interference pattern;

3 Lightning protection and grounding engineering construction shall meet the relevant requirements of Chapter 16 in this code.

14.4 System Debugging

14.4.1 In addition to the requirements of Section 6.4 in the current national standard GB 50348-2004 *Technical Code for Engineering of Security and Protection System*, the alarm system debugging shall also meet the following requirements:

1 According to the requirements of the current national standard GB 50394 *Code of Design for Intrusion Alarm Systems Engineering*, the functions and indexes such as the detection range, sensitivity, false alarm, alarm failure, restoration after alarming and anti-dismantle protection of the detector shall be inspected and the inspection result shall meet the design requirements;

2 The alarm linkage function, electronic map display function as well as the system response time from alarm to display and video recording shall be inspected and the inspection result shall meet the design requirements.

14.4.2 In addition to the requirements of Section 6.4 in the current national standard GB 50348-2004 *Technical Code for Engineering of Security and Protection System*, the video security system debugging shall also meet the following requirements:

1 The coordination, control and functional devices of camera and lens shall be inspected, which shall work normally and be free from obvious backlight phenomenon;

2 Clear and visible characters indicating the camera position, time and date shall be superimposed on the image display screen;

3 Storey mark shall be superimposed on the camera image screen in the elevator car; the elevator occupant images shall be clear;

4 When this system is integrated with other systems, the internet interface of this system and the integrated system as well as the centralized management and integrated control ability of this system shall be inspected;

5 Alarm function for video model loss shall be inspected;

6 The image reducibility and delay of digital video system shall meet the design requirements;

7 The word processing, dynamic alarm information processing, figure and image processing and operation of integrated security management system shall be completed on one set of computer system.

14.4.3 In addition to the requirements of Section 6.4 in the current national standard GB 50348-2004 *Technical Code for Engineering of Security and Protection System*, the access control system debugging shall also meet the following requirements:

1 As for every time effective entering, the system shall be stored the relevant information of the entering personnel and shall give off-site alarm for non-effective entering and forced entering;

2 The response time and event recording function of the system shall be inspected and the inspection result shall meet the design requirements;

3 When the system is arranged uniformly with one card for attendance, charging and target directing (garage), the system safety management shall meet the design requirements;

4 The linkage or integration function between the access control system and the alarm and guard tour systems shall be debugged. The linkage function between the access control system and fire automatic alarm system shall be debugged and the linkage and integration function shall meet the design requirements;

5 The internet interface between the system and intelligentized integration system shall be inspected, which shall meet the design requirements.

14.4.4 In addition to the requirements of Section 6.4 in the current national standard GB 50348-2004 *Technical Code for Engineering of Security and Protection System*, the visitor (video) intercom system debugging shall also meet the following requirements:

1 The image quality of the video intercom system shall meet the relevant requirements of the current professional standard GA/T 269 *Black-white Video Doorphone System*; the sound shall be clear and the sound level shall not be less than 80dB;

2 The two-way intercom function, remote control unlocking function, password unlocking function and standby battery of the system shall meet the relevant requirements of the current professional standard GA/T 72 *General Specifications of Building Intercom System and Elec-control Anti-burglary Door* and the design requirements.

14.4.5 In addition to the requirements of Section 6.4 in the current national standard GB 50348-2004 *Technical Code for Engineering of Security and Protection System*, the parking garage (lot) management system debugging shall also meet the following requirements:

1 The position and response speed of the induction coil shall meet the design requirements;

2 The signal indication, charging and security functions of the system for vehicles passing in/out shall meet the design requirements;

3 The equipment on the entrance and exit lanes shall work normally; the IC card reading/writing, display, lifting control of automatic gate machine, entrance/exit image information collecting as well as the real-time communication with the charging host shall meet the design requirements;

4 The parameter setting of charging management system, selling and loss reporting of IC card as well as data collection, statistics, summarizing and statement printing shall meet the design requirements.

14.4.6 The joint debugging, linkage and function integration of the system shall meet the following requirements:

1 The subsystems shall be inspected and debugged according to the system design requirements and the technical instruction of relevant equipment; each subsystem shall work normally;

2 After the alarm signal is simulated input, the linkage function of video monitoring system shall meet the design requirements;

3 The video monitoring system and access control system shall be linked with the fire automatic alarm system, and the linkage function shall meet the design requirements.

14.5 Self-examination and Test

14.5.1 In addition to the requirements of Article 8.3.4 in the current national standard GB 50339-2003 *Code for Acceptance of Quality of Intelligent Building Systems*, Chapter 7 in GB 50348-

2004 *Technical Code for Engineering of Security and Protection System* and GB 50395 *Code of Design for Video Monitoring System*, the inspection of video surveillance & control system shall also meet the following requirements:

1 Parameters such as real-time image quality, stored playback image quality, system time delay, time delay jitter, package loss rate of the video surveillance & control system shall be inspected and shall meet the requirements of Article 14.3.8 in the current professional standard JGJ 16-2008 *Code for Electrical Design of Civil Buildings* or the requirements of design document;

2 The linkage control function of the video surveillance & control system with the access control system, intrusion alarm systems, patrol management system and parking area (garage) management system shall be inspected and shall meet the design requirements;

3 The linkage control function of the video surveillance & control system with automatic fire alarm shall be inspected and shall meet the requirements of Article 13.4.7 in the current professional standard JGJ 16-2008 *Code for Electrical Design of Civil Buildings* or the requirements of design document.

14.5.2 In addition to complying with the requirements of Article 8.3.6 in the current national standard GB 50339-2003 *Code for Acceptance of Quality of Intelligent Building Systems*, the inspection of intrusion alarm system shall also include the inspection of image change alarm function, background change alarm function, behavior analysis and pattern recognition alarm function of the video alarm detector.

14.5.3 In addition to complying with the requirements of Article 8.3.7 in the current national standard GB 50339-2003 *Code for Acceptance of Quality of Intelligent Building Systems*, the inspection of access control system shall also include the inspection of recognition function, accuracy and linkage function of the biological recognition system and shall meet the requirements of Article 13.4.7 in the current professional standard JGJ 16-2008 *Code for Electrical Design of Civil Buildings* or the requirements of design document.

14.5.4 The inspection of patrol management system shall comply with the requirements of Article 8.3.8 in the current national standard GB 50339-2003 *Code for Acceptance of Quality of Intelligent Building Systems*.

14.5.5 The inspection of parking garage (lot) management system shall comply with the requirements of Article 8.3.9 in the current national standard GB 50339-2003 *Code for Acceptance of Quality of Intelligent Building Systems*.

14.5.6 The inspection of integrated security management system shall comply with the requirements of Article 8.3.10 in the current national standard GB 50339-2003 *Code for Acceptance of Quality of Intelligent Building Systems*.

14.6 Quality Record

14.6.1 In addition to complying with the requirements of Section 3.7 in this code, the quality record of security system shall also meet the relevant requirements of the current national standard GB 50348 *Technical Code for Engineering of Security and Protection System*.

15 Intelligentized Integration System

15.1 Construction Preparation

15.1.1 Technical preparation shall meet the following requirements:

1 According to the design document requirements and functional requirements, the construction organization shall complete the network planning, allocation scheme, integration system function and system performance document as well as the system linkage function demand table of the intelligentized integration system, which shall be approved through joint review;

2 The intelligentized integration system shall realize the following functions:

1) It shall be able to integrate the collection, conversion, storage, condition judgment, numerical calculation, graphical real-time display and integrated query of the integrated subsystem data;

2) It shall be able to realize the manual control as well as the automatic operation optimization control, timing control and energy-saving control of the integrated subsystem when the integrated subsystem may be controlled;

3) It shall be able to integrate the linkage control, authority management and emergency plan management of several subsystems;

4) It shall be able to integrate the operation fault, alarm prompt and processing of the subsystem and integration system;

5) It shall be able to realize information data sharing of integrated subsystems;

6) The intelligentized integration system shall not control the fire automatic alarm system and affect the independent operation of it;

7) It should be possessed of building energy consumption statistics, analysis and report functions and be able to provide energy consumption statistical data to the public building energy consumption monitoring system through international standard interface.

3 The communication interface and protocol of integrated subsystem shall meet the integration function and performance requirements; the physical interface should be adopted RS-232, RS-485, Ethernet and international standard interface;

4 As for the subsystem conducting real-time data collection and control, such materials as OPC server communication interface complying with OPC data access specification as well as subsystem OPC server parameter description and the test edition of OPC server software shall be provided;

5 As for the subsystem collecting historical operation records, multi-user database access interface complying with ODBC specification as well as subsystem database access interface description and database example (including test data) shall be provided;

6 The subsystem conducting video image collection and monitoring shall meet the following requirements:

1) Analog video matrix shall provide at least one circuit of analog composite video signal

port which shall be able to output all video images in sequence through switching;

  2) The analog video matrix shall provide communication port and its communication protocol; the communication protocol control command shall include input/output switching, camera control, PTZ control, preset control, etc. ;

  3) The digital video system shall provide software development kit in ActiveX control form, including such functions as real-time video display, video replay, video retrieval, input/output switching, camera control, PTZ control, preset control, photographing, video, etc. ,

  4) The equipment and software of digital video system shall be possessed of multi-user simultaneous access function.

 7 The communication protocol of integrated subsystem shall meet the following requirements:

  1) The communication protocol shall cover the description for such contents as data format, synchronisation mode, transmission speed, transmission procedure, error detection and correction method, authentication method, control character definition, function and shall also include examples;

  2) Serial communication protocol shall cover the description for such parameters as connection mode, baud rate, data bit, check bit and stop bit;

  3) Ethernet communication protocol shall cover the description for such parameters as transmission layer protocol, working mode and port number.

 8 The communication interface shall be subjected to function and performance test;

 9 Where the integration system is involved with the connection with more than two subsystems, mutual interference between systems shall be prevented.

15.1.2 The preparation of materials and equipment shall meet the following requirements:

 1 The equipment and software must be subjected to product quality inspection according to the requirements of Section 3.2 in the current national standard GB 50339 - 2003 *Code for Acceptance of Quality of Intelligent Building Systems* and shall meet the requirements of site acceptance;

 2 The technical documents provided for integrated subsystem shall meet the following requirements:

  1) The documents shall include the system diagram, network topology diagram, schematic diagram, plan, equipment parameter list, configuration monitoring interface file and editing software;

  2) The documents shall be in both paper and electronic forms; the document contents shall be consistent with the equipment and software installed on the project site;

  3) The document contents shall be consistent with the equipment parameter mark of the communication interface.

 3 The product information of integrated subsystem shall cover the following contents:

  1) System structure description, instruction manual and installation configuration manual;

  2) Integrated subsystem server and workstation software for test use;

  3) Instruction manual, installation configuration manual, development reference manual and connection description for the communication interface of integrated subsystem.

15.1.3 The integrated subsystem shall be possessed of the relevant acceptance conditions complying with the current national standard GB 50339 *Code for Acceptance of Quality of Intelligent Building Systems*.

15.2 Installation of Hardware and Software

15.2.1 System diagram, network topology diagram, equipment layout and wiring diagram shall be drawn according to the network planning and allocation scheme, integration system function and system performance documents.

15.2.2 The graphical interfaces shall be drawn and communication parameters shall be configured according to the technical documents of integrated subsystem; meanwhile, the subsystem authority management configuration shall be carried out.

15.2.3 The quality inspection of application software shall be carried out according to the integration system function and system performance documents, integrated subsystem communication interface, switching software of development communication interface as well as the requirements of Article 3.5.4 in this code.

15.2.4 Installation of such equipment as server, workstation, communication interface converter and media coder shall meet the requirements of Article 6.2.1 in this code.

15.2.5 The software installation of server and workstation shall meet the requirements of Article 6.2.2 in this code.

15.2.6 The software debugging and modification of communication interface shall be carried out on the dedicated computer and be subjected to version control.

15.2.7 The software at the service end of integration system shall be configured as automatic operation mode after startup.

15.3 Quality Control

15.3.1 The dominant items shall meet the following requirements:

 1 The hardwired connection and equipment interface connection of integrated subsystem shall meet the requirements of Article 10.3.6 in the current national standard GB 50339-2003 *Code for Acceptance of Quality of Intelligent Building Systems*;

 2 The software and equipment shall be free from runtime errors in startup, operation and shutdown processes;

 3 After the communication interface software has been modified, it shall pass the system test and regression test;

 4 The operation data shall be checked according to the communication interface, engineering information and actual equipment operating conditions of the integrated subsystem;

 5 The system shall be able to correctly realize the linkage function of intelligentized integration system as approved through joint review.

15.3.2 The general items shall meet the following requirements:

 1 The network address of such equipment as server, workstation, communication interface converter and media coder shall be configured according to the network planning and allocation scheme;

 2 The basic platform software and antivirus software of the operation system and database

shall be possessed of formal software use (authorization) license;

3 The operation system of the server and workstation shall be set in an auto update operation mode;

4 The server and workstation shall be equipped with antivirus software and be set in an auto update operation mode;

5 The configuration parameters of such equipment as server, workstation, communication interface converter and video codec shall be recorded.

15.4 System Debugging

15.4.1 The debugging preparation shall meet the following requirements:

1 The communication interface of integrated subsystem shall be completed installation;

2 The equipment and software of integration system shall be completed installation;

3 The graphical interface and parameters of integration system shall be completed configuration.

15.4.2 After the network parameters have been configured, the equipment and software of the integration system and subsystem shall be able to be connected to each other.

15.4.3 The software which requires continuous operation in the system debugging process shall always be in operation state.

15.4.4 The working state and operating log of the software shall be inspected every day and the error shall be corrected.

15.4.5 After the system has been debugged, the following inspections shall be carried out and errors be corrected:

1 The operation data collected by the integration system shall be compared with the actual operation data of the equipment;

2 Operation shall be carried out in the operation control interface of the integration system and be compared with the actual actuation of the equipment;

3 Historical data query shall be carried out on the integration system by using various query conditions and be compared with the corresponding historical data of the integrated subsystem;

4 The video monitoring image of the integration system shall be checked and compared with the actual output image of the picture pickup device.

15.4.6 After the data have been checked, functional test shall be carried out one by one according to the integration system function documents as approved through joint review.

15.4.7 After the functional test has been finished, performance test shall be carried out one by one according to the integration system function documents as approved through joint review.

15.4.8 If operation error occurs, or system function or performance fails to meet the design requirements in the debugging process, these shall be recorded completely, and the error shall be corrected and the function shall be enhanced.

15.4.9 Before the end of system debugging, all problem reports shall be treated and recorded.

15.5 Self-examination and Test

15.5.1 The system shall be inspected according to the integration system function and performance documents as approved through joint review; the system shall meet the requirements of documents.

15.5.2 The system network security shall be inspected according to the network planning and allocation scheme of intelligentized integration system and the system shall meet the requirements of documents.

15.5.3 The allocation scheme and configuration documentation of the equipment and software shall be complete.

15.5.4 After self-examination and test, all test users and test data shall be deleted. Before deleting, the test data shall be backuped.

15.6 Quality Record

15.6.1 Table B. 0. 22 in this code shall be filled in for the Linkage Function Demand List of Intelligentized Integration System.

15.6.2 Table B. 0. 23 in this code shall be filled in for the Equipment Parameter List of Integrated Subsystem.

15.6.3 Table B. 0. 24 in this code shall be filled in for the Communication Interface List of Integrated Subsystem.

15.6.4 Table B. 0. 25 in this code shall be filled in for the Network Planning and Configuration Table of Intelligentized Integration System.

16 Lightning Protection and Grounding

16.1 Equipment Installation

16.1.1 In addition to complying with the requirements of Section 6.2 in the current national standard GB 50343 - 2004 *Technical Code for Protection Against Lightning of Building Electronic Information System* and Chapter 24 in GB 50303 - 2002 *Code of Acceptance of Construction Quality of Electrical Installation in Building*, the installation of grounding body shall also meet the following requirements:

1 The vertical length of grounding body shall not be less than 2.5m and the spacing should not be less than 5m;

2 The buried depth of grounding body should not be less than 0.6m;

3 Distance between grounding body and building shall not be less than 1.5m.

16.1.2 In addition to complying with the requirements of Section 6.3 in the current national standard GB 50343 - 2004 *Technical Code for Protection Against Lightning of Building Electronic Information System* and Chapter 25 in GB 50303 - 2002 *Code of Acceptance of Construction Quality of Electrical Installation in Building*, the installation of grounding wire shall also meet the following requirements:

1 Where the main bar of building structure is used as grounding wire, it shall be welded to the main bar in the foundation and the number of welded shall be determined according to the dimension of main bar diameter and shall not be less than 2 roots of weld;

2 The grounding wire led to the grounding terminal shall be adopted multi-strands copper wire whose sectional area is not less than 4mm^2.

16.1.3 In addition to complying with the requirements of Section 6.4 in the current national standard GB 50343 - 2004 *Technical Code for Protection Against Lightning of Building Electronic Information System* and Chapter 27 in GB 50303 - 2002 *Code of Acceptance of Construction Quality of Electrical Installation in Building*, the equipotential connection installation shall also meet the following requirements:

1 The grounding wire of the main equipotential connecting terminal board of the building shall be directly led from the grounding device; the main equipotential bonding devices in each region shall be connected mutually;

2 Connecting conductors shall be led from two places of the grounding device to connect with the indoor main equipotential grounding terminal board; the sectional area of the connecting conductor of grounding device and indoor main equipotential grounding terminal board shall not be less than 50mm^2 for copper grounding wire and shall not be less than 80mm^2 for steel grounding wire;

3 The equipotential grounding terminal boards shall be connected by bolts; the copper grounding wire shall be welded or crimped and steel grounding wire shall be welded;

4 As for the grounding of each electrical equipment, independent grounding wire shall be adopted to connect with the grounding main line;

5 The plicate tube, metal sheath or metal mesh of the pipe insulation layer and the cable metal sheath shall not be used as grounding wire; the bridge and metal wire pipe shall not be used as grounding wire.

16.1.4 In addition to complying with the requirements of Section 6.5 in the current national standard GB 50343 – 2004 *Technical Code for Protection Against Lightning of Building Electronic Information System*, the installation of surge protective device shall also meet the following requirements:

1 Waterproof measures shall be provided for outdoor installation;

2 The installation position of surge protective device shall be close to the protected equipment.

16.1.5 In addition to complying with the requirements of Section 6.6 in the current national standard GB 50343 – 2004 *Technical Code for Protection Against Lightning of Building Electronic Information System*, Articles 3.4.6, 3.4.7 and 3.8.9 in GB 50169 – 2006 *Code for Construction and Acceptance of Grounding Connection Electric Equipment Installation Engineering* and Articles 12.1.1 and 14.1.1 in GB 50303 – 2002 *Code of Acceptance of Construction Quality of Electrical Installation in Building*, the lightning protection and grounding of comprehensive pipeline shall also meet the following requirements:

1 The connection positions between the metal bridge and grounding trunk line shall not be less than 2 positions;

2 The copper-cored grounding wires shall be bridged on both ends of the connection plate between the non-galvanized bridges and the sectional area shall not be less than 4mm^2;

3 The galvanized steel pipe shall be bridged by using dedicated grounding fastener; the bridging line shall be adopted copper-cored cord with sectional area no less than 4mm^2. Where the non-galvanized steel pipe is adopted with threaded connection, bridge ground wire shall be welded to both ends of the junction;

4 The shielding layer of armored cable shall be connected to the equipotential terminal block at the position entering the household.

16.1.6 The lightning protection and grounding of fire alarm and control system shall comply with the requirements of Article 5.4.7 in the current national standard GB 50343 – 2004 *Technical Code for Protection Against Lightning of Building Electronic Information System* and Section 3.11 in GB 50166 – 2007 *Code for Installation and Acceptance of Fire Alarm System*.

16.1.7 In addition to complying with the requirements of Article 5.4.6 in the current national standard GB 50343 – 2004 *Technical Code for Protection Against Lightning of Building Electronic Information System* and Article 6.3.6 in GB 50348 – 2004 *Technical Code for Engineering of Security and Protection System*, the lightning protection and grounding of security system shall also meet the following requirements:

1 The outdoor equipment shall be provided with lightning protection grounding and surge protective device for line;

2 The outdoor AC power line and control signal line shall be provided with metal shielding layer and be buried laying through steel pipe; both ends of the steel pipe shall be grounded reliably;

3 The outdoor video camera shall be arranged within the effective protection scope of lightning arrester or other air-termination conductor;

   4 The lightning ground resistance of the video camera pole grounding electrode shall be less than 10Ω;

   5 The equipment metal enclosure, cabinet, console, exposed metal pipe, trough, outer-layer of shielded wire cable and grounding terminal of surge protective device shall be connected to the grounding terminal of equipotential connection network in shortest distance.

16.1.8 The lightning protection and grounding of building automation system shall comply with the relevant requirements of the current national standard GB 50343 *Technical Code for Protection Against Lightning of Building Electronic Information System*.

16.1.9 The lightning protection and grounding of cable television system shall comply with the requirements of Article 5.4.8 in the current national standard GB 50343 - 2004 *Technical Code for Protection Against Lightning of Building Electronic Information System* and Article 6.3.6 in GB 50200 - 94 *Technical Code for Regulation of CATV System*.

16.1.10 The lightning protection and grounding of information facility system shall comply with the requirements of Article 5.4.4 in the current national standard GB 50343 - 2004 *Technical Code for Protection Against Lightning of Building Electronic Information System*.

16.1.11 The lightning protection and grounding of information network system shall comply with the requirements of Article 5.4.5 in the current national standard GB 50343 - 2004 *Technical Code for Protection Against Lightning of Building Electronic Information System*.

16.1.12 The lightning protection and grounding of broadcasting system shall comply with the requirements of Article 5.4.8 in the current national standard GB 50343 - 2004 *Technical Code for Protection Against Lightning of Building Electronic Information System*.

16.1.13 In addition to complying with the relevant requirements of the current national standard GB 50343 *Technical Code for Protection Against Lightning of Building Electronic Information System*, the lightning protection and grounding of generic cabling system shall also meet the following requirements:

   1 As for the cable entering into buildings, surge protective device shall be installed at the entry;

   2 Where the wire cable enters into the building, the metal sheath or metal part of electric cable and optical cable shall be connected to the nearby equipotential terminal board at the entry;

   3 The distribution cabinet (frame, box) shall be connected to the nearby equipotential device by insulated copper conductor;

   4 The equipment metal enclosure, cabinet, metal pipe, trough, outer-layer of shielded wire cable, static protective grounding of equipment, safety protection grounding, grounding terminal of surge protective device shall be connected to the grounding terminal of the nearby equipotential connection network.

16.2 Quality Control

16.2.1 Dominant items shall meet the relevant requirements of the current national standard GB 50339 *Code for Acceptance of Quality of Intelligent Building Systems*.

16.2.2 The general items shall meet the following requirements:

   1 As for the welding of steel grounding wire, the weld shall be full and anti-corrosive measures shall be taken;

 2 Metal sleeve shall be provided at the position where the grounding wire passes through the wall and floor-slab; and the metal sleeve shall be connected to the grounding wire.

16.3 System Debugging

16.3.1 After the lighting protection and grounding system has been installed, the ground resistance shall be tested and shall meet the requirements of Article 16.2.1 in this code.

16.3.2 After the equipotential bonding installation has been completed, conductivity test shall be carried out.

16.4 Self-examination and Test

16.4.1 The exposed parts of the grounding grid in equipotential bonding of the building shall be connected reliably; their specification shall be correct, paint is in good condition and marks are complete and obvious.

16.4.2 The inspection of grounding device shall meet the following requirements:

 1 The structure and installation position of grounding device shall be inspected;
 2 The buried spacing and depth of grounding body shall be inspected;
 3 The grounding resistance of grounding device shall be inspected.

16.4.3 The specification of grounding wire and its connection with the equipotential grounding terminal board shall be inspected.

16.4.4 The installation position, material specification and connection of equipotential grounding terminal board shall be inspected.

16.4.5 The parameter selection, installation position and connecting wire specification of surge protective device shall be inspected.

16.5 Quality Record

16.5.1 Table B.0.26 in this code shall be filled in for the Inspection Record of Electrical Grounding Device.

16.5.2 Table B.0.27 in this code shall be filled in for the Ground resistance Test Record.

17 Machine Room Engineering

17.1 Construction Preparation

17.1.1 The construction preparation shall not only comply with Section 3.3 in this code, but also meet the following requirements:

1 The layout and zoning of machine room shall meet the requirements of Section 23.2 in the current professional standard JGJ 16 - 2008 *Code for Electrical Design of Civil Buildings*;

2 The construction civil work profession of machine room shall be completed, the ground shall be levelled and cleaned, and the requirements of Article 23.3.2 in the current professional standard JGJ 16 - 2008 *Code for Electrical Design of Civil Buildings* shall be met;

3 The installation for water supply and drainage pipelines in the machine room shall be free from leakage.

17.2 Equipment Installation

17.2.1 In addition to complying with the requirements of Chapter 10 in the current national standard GB 50462 - 2008 *Code for Construction and Acceptance of Electronic Information System Room*, the construction of the indoor decoration engineering of machine room shall also meet the following requirements:

1 Decoration engineering may be carried out after the lightning grounding equipotential board has been installed, cabinet wire slot has been led in and pipelines have been installed;

2 The movable floor support frame shall be installed firmly and be leveled;

3 The movable floor height shall be determined according to cable wiring and the air supply requirements of air conditioning, and it should be 200mm~500mm;

4 The floor wire cable outlet shall be located according to the actual position of the computer and outlet shall be provided with wire cable protective measures.

17.2.2 In addition to complying with the requirements of Chapter 3 in the national standard GB 50462 - 2008 *Code for Construction and Acceptance of Electronic Information System Room*, the construction of the power supply and distribution system engineering of machine room shall also meet the following requirements:

1 The fabrication dimension of the mounting bracket of distribution cabinet and box shall match with the dimension of the distribution cabinet and box; the mounting bracket shall be installed firmly and grounded reliably;

2 The construction of wire slot, wire pipe and wire cable shall meet the requirements of Chapter 4 in this code;

3 The installation of luminaries, switches, various electrical control gears and various sockets shall meet the following requirements:

    1) The luminaries, switches and sockets shall be installed firmly at accurate position; the switch position shall correspond to the lamp position;

    2) The socket panels at the sample plane height in the same room shall be horizontal;

3) The determination of the support, hanger and fixing point positions of luminaries shall meet the principle of firmness and safety, tidiness and aesthetics;

4) After the luminaries and distribution box have been installed, the insulation of each branch line shall be measured; the insulation resistance shall be greater than 1MΩ and be recorded;

5) The machine room floor slab shall meet the load-bearing requirements of the battery set;

4 The installation of uninterruptible power supplies shall meet the following requirements:

1) The main machine and battery cabinet shall be fixed according to the design requirements and product technical requirements;

2) The wiring of various wire cables shall be firm and correct and be marked;

3) The battery set of uninterruptible power supply shall be grounded by DC grounding mode.

17.2.3 The construction of lightning protection and grounding system engineering shall comply with the requirements of Chapter 4 in the current national standard GB 50462 - 2008 *Code for Construction and Acceptance of Electronic Information System Room* and Chapter 16 in this code.

17.2.4 The construction of generic cabling system engineering shall comply with the requirements of Chapter 7 in the current national standard GB 50462 - 2008 *Code for Construction and Acceptance of Electronic Information System Room* and Chapter 5 in this code.

17.2.5 The construction of security system engineering shall comply with the requirements of Chapter 8 in the current national standard GB 50462 - 2008 *Code for Construction and Acceptance of Electronic Information System Room* and Chapter 14 in this code.

17.2.6 The construction of air conditioning system engineering shall comply with the requirements of Chapter 5 in the current national standard GB 50462 - 2008 *Code for Construction and Acceptance of Electronic Information System Room*.

17.2.7 The construction of water supply and drainage system engineering shall comply with the requirements of Chapter 6 in the current national standard GB 50462 - 2008 *Code for Construction and Acceptance of Electronic Information System Room*.

17.2.8 The construction of electromagnetic shielding engineering shall comply with the requirements of Chapter 10 in the current national standard GB 50462 - 2008 *Code for Construction and Acceptance of Electronic Information System Room*.

17.2.9 The construction of fire-fighting system engineering shall comply with the relevant requirements of the current national standard GB 50263 *Code for Installation and Acceptance of Gas Extinguishing Systems*, Chapter 9 in GB 50462 - 2008 *Code for Construction and Acceptance of Electronic Information System Room* and Chapter 13 in this code.

17.2.10 The construction of secrete-involved network computer room shall meet the relevant national technical requirements for the gradational protection of information system involving state secrets.

17.3 Quality Control

17.3.1 The dominant items shall meet the following requirements:

1 The electrical device shall be installed firmly in order; its marks shall be identifiable and both its inside and outside be clean;

2 The anti-static construction of the ground and movable floor slab in the machine room shall meet the requirements of Section 23.2 in the current professional standard JGJ 16-2008 *Code for Electrical Design of Civil Buildings*;

3 The surge protective devices at the inlets of power supply line and signal line shall be installed firmly at correct position;

4 The grounding wire and equipotential grounding terminal board shall be connected correctly and installed firmly. The grounding resistance shall meet the requirements of 16.4.1 in this code.

17.3.2 The general items shall meet the following requirements:

1 The electrical devices in the suspended ceiling shall be installed at the place convenient for maintenance;

2 The power distribution unit shall be provided with clear sign indicating capacity, voltage and frequency, etc.;

3 The pedestal of landing type electrical device shall be firmly installed on the floor;

4 The power supply line and signal line shall be laid separately, arranged in order and fixed by bundling; their length shall be reserved with allowance;

5 The luminaries arranged in rows shall be straight, even and in order.

17.4 System Debugging

17.4.1 The debugging of generic cabling system shall comply with the requirements of Chapter 7 in the current national standard GB 50462-2008 *Code for Construction and Acceptance of Electronic Information System Room* and Chapter 5 in this code.

17.4.2 The debugging of security system shall comply with the requirements of Chapter 8 in the current national standard GB 50462-2008 *Code for Construction and Acceptance of Electronic Information System Room* and Chapter 14 in this code.

17.4.3 The debugging of air conditioning system shall comply with the requirements of Chapter 5 in the current national standard GB 50462-2008 *Code for Construction and Acceptance of Electronic Information System Room*.

17.4.4 The debugging of fire-fighting system shall comply with the requirements of Chapter 9 in the current national standard GB 50462-2008 *Code for Construction and Acceptance of Electronic Information System Room*, GB 50263 *Code for Installation and Acceptance of Gas Extinguishing Systems* and Chapter 13 in this code.

17.5 Self-examination and Test

17.5.1 The air conditioning environment in the machine room shall meet the requirements of Article 12.2.2 in the current national standard GB 50339-2003 *Code for Acceptance of Quality of Intelligent Building Systems*.

17.5.2 The noise inspection shall meet the following requirements:

1 The test points shall be arranged at the main operator position of 1.2m~1.5m away from the ground;

2 The machine room shall be kept away from the noise source; where it is inevitable, noise elimination and insulation measures shall be taken;

3 Equipment with strong noise should not be arranged in the machine room; where it is inevitable, effective sound insulation measures shall be taken; the noise value in the machine room should be 35dBA~40dBA.

17.5.3 The inspection of the power supply and distribution system shall meet the following requirements:

1 The voltage, frequency and waveform distortion rate shall be measured at the output end of distribution cabinet (board);

2 The power quality of power supply shall meet the requirements of Section 3.4 in the current professional standard JGJ 16 - 2008 *Code for Electrical Design of Civil Buildings*.

17.5.4 The illuminance inspection shall meet the following requirements:

1 The test points shall be arranged with a spacing of 2m~4m and shall be kept a distance of 1m to wall surface and 0.8m to the ground;

2 The machine room illuminance shall comply with the relevant requirements of the current national standard GB 50034 *Standard for Lighting Design of Buildings*.

17.5.5 The electromagnetic shielding inspection shall meet the following requirements:

1 Where the frequency is 0.15MHz~1,000MHz, the radio interference field intensity shall not be larger than 126dB.

2 The interference field intensity of magnetic field shall not be larger than 800A/m.

17.5.6 The grounding of machine room engineering shall meet the requirements of Article 23.4.2 in the current professional standard JGJ 16 - 2008 *Code for Electrical Design of Civil Buildings*. The inspection of ground resistance shall meet the requirements of Article 16.2.1 in this code.

17.6 Quality Record

17.6.1 The machine room engineering quality record shall not only comply with the requirements of Section 3.7 in this code, but also comply with the relevant requirements of the current national standard GB 50462 *Code for Construction and Acceptance of Electronic Information System Room*.

Appendix A Records of Project Implementation and Quality Control

Table A.0.1 Unpacking Inspection Record Sheet of Equipment

No.:

Equipment name			Inspection date	
Specification and type			Total quantity	
Packing list No.			Inspection quantity	
Inspection record	Packing condition			
	Accompanying documents			
	Spare parts and accessories			
	Appearance condition			
	Test condition			
Inspection result	Detailed list of defective attachments and spare parts					
	No.	Name	Specification	Organization	Quantity	Remarks

Conclusions:

Signature column	Employer (supervision) organization	Construction organization	Supply equipment organization

Note: This table is filled in and preserved by the construction organization.

Table A.0.2 Notice Sheet for Design Change

No.:

Engineering name			Profession name	
Name of design organization			Date	
No.	Figure No.		Changed contents	
Signature column	Emplpoyer (supervision) organization	Design organization		Construction organization

Note:

1 This table is respectively preserved of one part by the employer organization, supervision organization, construction organization and urban construction archives.

2 The one involving drawing alteration must be noted with the figure number of the drawing to be changed.

3 The design changes of different professions must not be transacted on the same notice sheet for change.

4 The column of "Profession name" shall be filled in accordance with professions, for example, building, structure, water supply and drainage, electrical, ventilation air conditioner, intelligent building engineering, etc.

Table A.0.3 Engineering Negotiation Record

No.:

Engineering name			Profession name	
Name of the proposing organization			Date	
Abstract				
No.	Figure No.		Negotiation content	

Signature column	Employer organization	Supervision organization	Design organization	Construction organization

Note:
1. This table is respectively preserved of one part by the employer organization, supervision organization, construction organization and urban construction archives.
2. The one involving drawing alteration must be noted with the figure number of the drawing to be changed.
3. The engineering negotiations of different professions must not be transacted on the same notice sheet for change.
4. The column of "Profession name" shall be filled in accordance with professions, for example, building, structure, water supply and drainage, electrical, ventilation air conditioner, intelligent building engineering, etc.

Table A.0.4 Joint Review Records of Drawings

No.:

Engineering name			Date	
Place			Profession name	
No.	Figure No.	Drawing problems	Drawing problem clarification	
Signature column	Employer organization	Supervision organization	Design organization	Construction organization

Note:
1. This table is sorted and summarized by the construction organization, and respectively preserved of one part by the employer organization, supervision organization, construction organization and urban construction archives.
2. The joint review records of drawings shall be summarized and sorted according to professions (building, structure, water supply and drainage, heating, electrical, ventilation air conditioner, intelligent system, etc.).
3. The design organization shall appoint professional design principal to sign the record, and other organizations concerned shall appoint the project technical director or the relevant profession director to sign and confirm the record.

Table A.0.5 Network Equipment Configuration Sheet

No.:

Engineering name		Inspection position		Date	
Construction organization		Registered construction engineer		Drawing number	

No.	Equipment name	Type	Placed position	Network segment partition	IP Address	Mask	Priority level	Parameter 1	Parameter 2

Table A.0.6　Application Software System Configuration Sheet

No. :

Engineering name				
Construction organization			Professional engineer	
Name and serial number of executive standards for construction			Number of design drawings	
Software system name			Version	

No.	Recorded items		Recorded content	Remarks
1	Equipment used in application software system	Equipment model		
		Installation position		
		Hardware configuration parameters		
		Software installation position		
2	Equipment network parameter	Physical address		
		IP address		
		Mask		
3	Equipment software platform	Operating system software		
		Database software		
		Antivirus software		
4	Installation instructions for modules of application software system			
5	Other descriptions:			

Appendix B Test Records

Table B.0.1 Pretest Record Sheet

No. :

Engineering name		Pretest items	
Pretest position		Inspection date	

Criteria: construction drawings (construction drawing number _____), Design change/negotiation (number _____) and relevant specifications and codes. Main material or equipment: _____ Specification/model: _____
Pretest contents: Declarant:
Inspection suggestions:
Review suggestions: Reviewer: Review date:

Construction organization		
Profession technical director:	Professional quality inspector	Professional headman

Note: This table is filled in and preserved by the construction organization.

Table B.0.2 Inspection Lot Test Records

No.:

Engineering name			Inspection position	
Construction organization			Registered construction engineer	

		Provisions of the construction quality acceptance specification	Inspection and evaluation records of construction organization	Acceptance records of supervision (employer) organization
Dominant item	1			
	2			
	3			
	4			
	5			
	6			
	7			
	8			
	9			
General item	1			
	2			
	3			
	4			
	5			
	6			
	7			
	8			
	9			
Inspection and evaluation results of the construction organization	Professional headman (builder)		Construction team leader	
	Professional project quality inspector: Date			
Acceptance conclusion of supervision (employer) organization	Agreed to acceptance Professional supervision engineer (Project technical director of employer organization): Date			

Table B. 0. 3 System Debugging Reports

No.:

Engineering name		System name	
Employer organization		Construction organization	
Registered construction engineer		Debugging date	

No.	Debugging content	Debugging result

Debugging condition	
Debugging personnel (sign)		Supervision engineer (sign)	
Construction organization (sign)		Design organization (sign)	

Table B.0.4 Network System Debugging Record Sheet

No.:

Engineering name			Acceptance position		
Construction organization			Professional headman		
Name and serial number of executive standard for construction			Design drawing (change) number		

No.	Equipment name	Model	Placed position	Network segment partition	IP address	Connected design	Connected inspection conditions	Remarks

Date:

Table B.0.5 Optical Node (Forward Direction) Debugging Record Sheet

No. :

Optical node number		Optical workstation model		Installation position	
Number of output port		AC voltage		DC voltage	
Design organization		Designer		Test personnel	
Test date		Test instrument		Test ambient temperature	

Input optical power: dBm		Detection DC voltage of optical received power: V				Relevant referential technical index			
Test frequency (MHz)	Optical module output electricity (dBμV)	Optical workstation downstream design value/debugging value (dBμV)				CNR (dB)	HUM (%)	MER (dB)	BER
		Port 1	Port 2	Port 3	Port 4				

Output slope (dB)	Front-stage		Remarks:
	Inter-stage		
Attenuator (dB)			
Equalizer (dB)			

Table B.0.6 Optical Node (Reverse Direction) Debugging Record Sheet

No. :

Optical node number				Optical workstation model				Installation position		
Number of output port				AC voltage				DC voltage		
Design organization				Designer				Test personnel		
Test date				Test instrument				Test ambient temperature		
Place of signal injection:				Optical power voltage of light emitting module: V				Optical received power voltage in machine room: V		Reference index
Frequency (MHz)	Injection level (dBμV)	Optical workstation upstream design value/debugging value (dBμV)				Optical transceiver output level in machine room (0dB decrease)		Optical transceiver output level in machine room (debugging value)		Gain inequality of transmission route in upstream (Gd)
		Port 1	Port 2	Port 3	Port 4	dBμV		dBμV		
Attenuator in reverse direction (dB)				Instruction: 1 The signal may be respectively injected from the optical station and the user side, and then debugged. 2 The real-time noise spectrum records of the output monitoring port may be used to test the upstream monitoring port of the optical workstation and the monitoring port of the optical transceiver in machine room (the same below).						
drive level of upstream light emitting module (dBμV)										
RF output level of optical transceiver in machine room (dBμV)										Gd test specification:
Real-time noise spectrum records of output monitoring port:				The maximum maintaining noise spectrum records at output monitoring port in 30s:				The maximum maintaining noise spectrum records at output monitoring port in 60s:		

Table B.0.7 Amplifier (Forward Direction) Debugging Record Sheet

No. :

Optical node number		Amplifier model		Test instrument	
Number of output port		Design organization		Designer	
Test personnel		Test date		Test ambient temperature	
Installation position					

Test frequency (MHz)	Input level (dBμV)	Amplifier output design value/debugging value (dBμV)				Relevant referential technical index			
		Port 1	Port 2	Port 3	Port 4	CNR (dB)	HUM (%)	MER (dB)	BER

Output slope (dB)		
Attenuator (dB)	Front-stage	
	Inter-stage	
Equalizer (dB)	Front-stage	
	Inter-stage	

Remarks:

• 103 •

Table B.0.8 Amplifier (Reverse Direction) Debugging Record Sheet

No. :

Optical node number		Amplifier model		Test instrument	
Number of output port		Design organization		Designer	
Test personnel		Test date		Test ambient temperature	
Installation position					

Place of signal injection:

Frequency (MHz)	Injection level (dBμV)	Amplifier upstream test value (dBμV)				Optical workstation upstream test value (dBμV)			
		Design value/debugging value				Design value/debugging value			
		Port 1	Port 2	Port 3	Port 4	Port 1	Port 2	Port 3	Port 4
Output slope (dB)									
Attenuator in reverse direction (dB)									
Equalizer in reverse direction (dB)									
Real-time noise spectrum records of output monitoring port :		The maximum maintaining noise spectrum records at output monitoring port in 30s:				The maximum maintaining noise spectrum records at output monitoring port in 60s:			

Table B.0.9 Front-end Equipment Debugging Record Sheet

No. :

Item	Channel	Direct receive and transmit							Frequency modulation broadcasting			Satellite receiving			Self-produced program	
		CH	CH	CH	CH	CH	CH	CH	MHz	MHz	MHz	CH	CH	CH	CH	CH
Level value	Front-end input level															
	Intermediate frequency output level															
	Demodulation output level															
Signal handling equipment	Satellite receiving output level															
	Modulation input level															
	Channel conversion input level															
	Digital modulator output level															
	Output channel															
	Front-end output level															
	Attenuator step															
Test time				Climate				Model of measuring instrument					Test personnel			

Table B.0.10 Record Sheet for Test Data of User Terminal

No.:

Construction organization			Design organization		
Test personnel			Test instrument		
Test date			Test ambient temperature		

Test frequency (MHz)	Level value at the test place (dBμV)						
Downstream							
Upstream							

Frequency/user side injection level (MHz/dBμV)	Amplifier upstream/machine room upstream input signal detecting port level value (link attenuation = user side injection level - detection level -20 dB)						

| Test frequency (MHz) | Test index | | | | | | |
	CNR (dB)	HUM (%)	C/CSO (dB)	C/CTB (dB)	MER (dB)	BER	Remarks

Table B.0.11 Broadcasting System Engineering Electroacoustic Performance Measurement Record Sheet

No.:

Measuring place	
Measuring instrument	
Measuring personnel	

Ensured sound pressure level, sound field non-uniformity and transmission frequency response measuring data							
Sound Pressure value (dB) / Measuring point / Center Frequency (Hz)	1	2	3	4	...	n	
80							
100							
125							
160							
200							
250							
315							
400							
500							
630							
800							
1k							
1.25k							
1.6k							
2k							
2.5k							
3.15k							
4k							
5k							
6.3k							
8k							
10k							
12.5k							
Overall sound pressure level (Flat)							

Leak out sound attenuation measuring data				
Measuring point	East	South	West	North
DB value				

Measuring data of speech transmission index of sound reinforcement system shall be recorded according to measuring methods in STIPA

Electroacooustic performance measuring result	Item	Ensured sound pressure level	Sound field non-uniformity	Leak out acoustic attenuation	Speech transmission index STIPA of sound enlargement system	System signal-to-noise ratio	Transmission frequency response
	Grade evaluation						

Record filling personnel	(Signature)	Date:
Record auditor	(Signature)	Date:

Table B. 0. 12 Quality Acceptance Record Sheet for Telephone Switching System

No. :

Engineering name of unit (subunit):			Sub-subsection engineering		
Name of subitem engineering			Acceptance position		
Construction organization			Registered construction engineer		
Name and serial number of the executive standard					
Subcontractor			Subcontract project manager		
	Inspection item (dominant item)			Inspection and evaluation record	Remarks
1	Inspection before power on		The nominal working voltage is −48V		The allowable variation scope is −57V ~ −40V
2	Hardware inspection and test		Visible and audible alarm shall work normally		
			Load the test procedure, confirm that the hardware system is without failure by self-examination		
3	System inspection and test		Various calls, maintenance management, signal aspect and network support functions of the system		
4	Preliminary inspection and test	Reliability	Shall not result in more than 50% of the subscriber lines and junction line failing in call		Meet the requirements of YD 5077
			Call drops or stop connection of each user group shall not be large than 0.1 time/ every month		
			Call drops or stop connection of the relaying group: 0.15 times/month (less than or equal to 64 speech path); 0.1 times/ month (64 speech path ~ 480 speech path)		
			Abnormal incoming call and outgoing call connection of the individual users: per thousand users shall be less than or equal to 0.5 users · times/month; per hundred relaying shall be less than or equal to 0.5 line · times/month		
			The restart index of the processor within one month shall be 1~5 times (including three kinds of restart)		
			Software test fault shall not greater than 8 pcs/ month, hardware as printed circuit board replacing times shall not greater than 0.05 times/100 household and 0.005 times/30 PCM systems per month		
			For a long time conversation, 12 intercoms shall maintain 48h		

Table B.0.12 (continued)

	Inspection item (dominant item)			Inspection and evaluation record	Remarks
4	Preliminary inspection and test	\multicolumn{2}{l}{Obstacle rate: the obstacle rate in the office shall not be greater than 3.4×10^{-4}}		40 users conduct simulated call for 100,000 times simultaneously
		Performance Test	Home exchange call		3 times to five times for sampling measurement each time
			Outgoing and incoming exchange call		Relaying 100% test
			Tandem trunking test (various modes)		Five times for sampling measurement of each mode
			Other types of exchange call		
			The index of charging error rate shall not exceed 10^{-4}		
			Special service business (extremely for 110, 119 and 120 etc.)		100% test
			The subscriber line is accessed into the modem and the transmission rate is 2400 bps, and the data error rate shall not be greater than 1×10^{-5}		
			2B+D user test		
		\multicolumn{2}{l}{Relaying test: the repeat circuit call test, take 2~3 circuits for sampling measurement (including various ringing conditions)}		Mainly for signaling and interfaces
		Call completing rate test	The call completing rate between offices shall up to more than 99.96%		60 pairs of users, 100,000 times
			The call completing rate between offices shall up to more than 98%		Call 200 times
		\multicolumn{2}{l}{Adopt man-machine command to carry out with failure diagnosis test}		

Inspection suggestions:

Supervision engineer signs (professional technical project leader of the employer organization):   Date:
Principal of test organization signs: Date:

Table B. 0. 13 Access Network Equipment Quality Acceptance Record Sheet

No. :

Engineering name of unit (subunit):			Sub-subsection project	
Subitem engineering name			Acceptance position	
Construction organization			Registered construction engineer	
Name and serial number of executive standards for construction				
Subcontractor			Subcontract project manager	

	Inspection item (dominant item)		Inspection and evaluation record	Remarks
1	Installation environment inspection	Machine room environment		The one meets the design requirements is regarded as qualified
		Power supply		
		Ground resistance value		
2	Equipment installation inspection	Pipeline laying		The one meets the design requirements is regarded as qualified
		Equipment cabinet and module		
3	Line interfaces of the transceiver	Power spectral density		The one meets the design requirements is regarded as qualified
		Longitudinal balance loss		
		Overvoltage protection		
	User network interface	25.6 Mbit/s electrical interface		
		10 BASE-T interface		
		USB interface		
		PCI interface		
	Service node interface	STM -1 (155 Mbit/s) optical interface		
		Telecommunication interface		
	Separator test			
	Transmission performance test			
	Functional verification test	Transmission function		
		Management function		

Inspection suggestions:

Supervision engineer signs (professional technical project leader of the employer organization):   Date:
Principal of test organization signs: Date:

Table B.0.14 Quality Acceptance Record Sheet for Clock System

No.:

Engineering name of unit (subunit):				Sub-subsection project		
Sub-item name				Acceptance position		
Construction organization				Registered construction engineer		
Name and serial number of executive standards for construction						
Subcontractor				Subcontract project manager		
Inspection item					Inspection records	Remarks
Dominant Project	1	Time information equipment working condition	GPS time service and time server			Time control of time information equipment, master clock and slave clock must be accurate and synchronous
			Time control and synchronization of systematic master clock			
			Systematic slave clock			
			Synchronization of system time			
	2	Electrical installation	Installation of antenna			Construction of system installation shall meet the requirements of the current national standard "Code of Acceptance of Construction Quality of Electrical Installation in Building" (GB 50303)
			Installation of outdoor display equipment			
			Installation of indoor display equipment			
			Installation of power supply			
			Wiring system			
			Lightning grounding			
General item	3	System test function	Time server			Monitor the operation conditions of systematic master clock, slave clock, time server, time service, etc.
			Systematic master clock			
			Systematic slave clock			
			Synchronization of system time			
	4	Control function	The master clock shall be synchronous with the receiver of timing signal, and the master clock shall conduct timing over the slave clock synchronously			The master clock shall be synchronous with the receiver of timing signal, and the master clock shall conduct timing over the slave clock synchronously
			The master clock shall conduct timing over the slave clock synchronously			
			Access control			
	5	Automatic recovery function	Timing automatic recovery after power-failure			The system shall have the function of automatic recovery after the power-failure
			Time service automatic recovery after power-failure			
	6	Time service function	Cover of the time service function			The clock system shall be possessed of the timing and time service function towards host machines of other weak current systems;
			Timing and time service function			
	7	System configuration	Independent timing accuracy of the master clock			Main technical parameters as the independent timing accuracy of master clock and the error of the master and slave clocks are synchronous
			Synchronization error of the master and slave clocks			
			Software system update			
Inspection suggestions:						
Supervision engineer signs (professional technical project leader of the employer organization):						
Principal of the test organization signs:						
Date:						

Table B.0.15 Quality Acceptance Record Sheet for Information Guidance and Release System

No. :

Engineering name of unit (subunit):		Sub-subsection project	
Sub-item name		Acceptance position	
Construction organization		Registered constructorcon engineer	
Name and serial number of executive standards for construction			
Subcontractor		Subcontract project manager	

Inspection item				Inspection records	Remarks
Dominant item	1	Installation of multi-media display screen	Installation of outdoor display equipment		The installation of multi-media display screen must be firm, the power supply and communication transfer system must be reliably connected to ensure the application requirements.
			Installation of indoor display equipment		
			Outdoor environment recovery		
	2	Electrical installation	Power supply		System installation shall meet the requirements of the current national standard GB 50303 *Code of Acceptance of Construction Quality of Electrical Installation in Building*
			Wiring system		
			Lightning grounding		
General item	3	System service function	Material management and edit		Treatment function and communication function shall meet the design requirements
			Broadcasting management and control		
			Communication of the system and each display screen		

Table B.0.15(continued)

Inspection item				Inspection records	Remarks
General item	4	System control function	Correctly display the content to be released		
			Supervision of the released effect		
	5	Automatic recovery function	Broadcasting automatic recovery after power-failure		
	6	Screen display inspection	Display layout inspection of information content		
			Display screen luminance and color inspection		
			Broadcasting quality inspection of sound and video		
	7	System configuration	System configuration management and log management		
			24h function and performance test		
			Harmony of display screen and environment		
			Software system update		
Inspection suggestions:					
Supervision engineer signs (professional technical project leader of the employer organization):					
Principal of the test organization signs:					
Date:					

Table B.0.16 Quality Acceptance Record Sheet for Call and Intercom System

No. :

Engineering name of unit (subunit):		Sub-subsection project	
Sub-item name		Acceptance position	
Construction organization		Registered construction engineer	
Name and serial number of executive standards for construction			
Subcontractor		Subcontract project manager	

		Inspection item		Inspection records	Remarks
Dominant item	1	Inspection of call and intercom	Response of host machine and each terminal (encoding correspondence)		The system shall respond the call in time and correctly, and the image and voice shall be distinct
			Whether the response is in time		
			Image quality inspection		
			Sound quality inspection		
	2	Access guard control inspection	Inspection of corresponding sheet of the access guard		System installation shall meet the requirements of the current national standard GB 50303 *Code of Acceptance of Construction Quality of Electrical Installation in Building*
			Whether each call is possessed of timely and correct response		
			Access guard installation inspection		
			Access guard response inspection		
General item	3	System service function	Inspection of call and intercom		Treatment function and communication function shall meet the design requirements
			Paging function inspection		
			Broadcasting function inspection		
			Broadcasting management and control inspection		

Table B. 0. 16(continued)

Inspection item				Inspection records	Remarks
General item	4	Electrical installation	Installation of host machine system		
			Installation of terminal		
			Display screen installation		
			Installation of broadcasting equipment		
			Lightning grounding		
			Line wiring inspection		
	5	Terminal display inspection	Terminal image quality inspection		
			Terminal sound quality inspection		
	6	System configuration	System configuration management and log management		
			24h function and performance test		
			Software system update		

Inspection suggestions:

Supervision engineer signs: Principal of test organization signs:

(The project professional technical director of employer organization)

Date: Date:

Table B.0.17 Quality Acceptance Record Sheet for Ticket Selling and Checking System

No. :

Engineering name of unit (subunit):		Sub-subsection project	Information facility system
Sub-item name	Ticket selling and checking system	Acceptance position	
Construction organization		Registered construction engineer	
Name and serial number of executive standards for construction			
Subcontractor		Subcontract project manager	

Inspection item				Inspection records	Remarks
Dominant item	1	Ticket selling function inspection	Ticket selling function		Function test of the ticket selling process of the ticket-selling machine
			Card making function		
			Settlement function		
	2	Ticket management inspection	Statistics of ticket selling data		Statistical accuracy of ticket selling data and checking data shall be carried out with concurrent data simulation test and inspection
			Statistics of ticket checking data		
			Concurrent data simulation		
General item	3	Inspection of ticket checking gate machine	Response to the ticket checking result		Carry out with simulating ticket checking towards each equipment respectively with a ticket checking gate machine
			Gate machine opening effect		
	4	System installation inspection	Ticket-selling machine installation		System installation shall meet the requirements of the current national standard GB 50303 *Code of Acceptance of Construction Quality of Electrical Installation in Building*
			Installation of ticket checking gate machine		
			Network and computer equipment installation		
			Lightning grounding		

Table B.0.17(continued)

		Inspection item		Inspection records	Remarks
General item	5	System installation inspection	Guide the installation of the guard rail		Treatment function and communication function shall meet the design requirements
			Installation of ticket selling and checking terminal equipment		
			Installation of power supply		
			Wiring system		
			Emergency backup function inspection		
	6	System control and service function	Communication transmission function		
			System self-examination and test function		
	7	Automatic recovery function	Automatic recovery function after power-failure		
	8	Screen display inspection of ticket selling machine	Display layout inspection of information content		
			Display screen luminance and color inspection		
	9	System configuration	System configuration management and log management		
			24h function and performance test		
			Software system update		

Inspection suggestions:

Supervision engineer (professional technical project leader of employer organization) signs:

Principal of the test organization signs:

Date:

Table B. 0. 18 Function Table of information application system

No. :

System (engineering) name		Construction organization		
System function instruction				
No.	Function category	Function name	Detailed description	Remarks

Employer organization	User's organization	Supervision organization	Construction organization
Principal Date: Seal:	Principal Date: Seal:	Principal Date: Seal:	Principal Date: Seal:

Table B.0.19 Configuration Parameter Record Sheet of the Information Application System

No.:

System (engineering) name		Construction organization		
No.	Recorded item	Recorded content		Remarks
1	Equipment number			
2	Equipment usage			
3	Specification and model			
4	Hardware configuration parameter			
5	Installation position			
6	Network parameter	Physical address		
		IP address		
7	Operating system software	Software version		
		Installation position		
		User name and password of the administrator		
8	Database software	Software version		
		Installation position		
		User name and password of the administrator		
9	Antivirus software	Software version		
		Installation position		
10	Network firewall software	Software version		
		Installation position		
11	Application system software	Software version		
		Installation position		
		User name and password of the administrator		
		Configuration parameter		
12	Other parameters			
Signature of the note-taker:		Supervision engineer (or employer organization) signs:		Record date:

Table B.0.20 Control Unit Cable Test Record

No. :

System (engineering) name			Construction organization		
Test date			Instrumentation		
Control box terminals	No.	Terminal equipment	Model	Continuity check of the line	Remarks
Conclusion					
Employer (supervision) organization representative			Quality inspector		
Profession technical director:			Test personnel		
Operator					

Table B. 0. 21 Single-point Debugging Record Sheet

No.:

System (engineering) name			Construction organization			
Debugging date			DDC box numbering		Instrumentation	

No.	Description	No.	Point name	Type	Point address	Terminal equipment	Pass/No pass
1							
2							
3							
4							
5							
6							
7							
8							
9							
10							
11							
12							
13							
14							
15							
16							
17							
18							
19							
20							

Employer (supervision) organization representative		Quality inspector	
Profession technical director		Test personnel	

Table B.0.22 Linkage Function Demand List of Intelligentized Integration System

No.:

System (engineering) name		Construction organization			
Cross-subsystem linkage function demand instruction					
Linkage triggering conditions	Actuation of linkage			Linkage function usage instruction	Remarks
Subsystem name	Subsystem name	Control item name	Actuation		
Name of parameter item					
Triggering conditions					
Subsystem name	Subsystem name	Control item name	Actuation		
Name of parameter item					
Triggering conditions					
Subsystem name	Subsystem name	Control item name	Actuation		
Name of parameter item					
Triggering conditions					
Subsystem name	Subsystem name	Control item name	Actuation		
Name of parameter item					
Triggering conditions					
Employer Development organization	User's organization	Design organization	Supervision organization	Construction organization	
Principal: Date: Seal:	Principal: Date: Seal:	Principal: Date: Seal:	Principal: Date: Seal:	Principal: Date: Seal:	

Table B.0.23 Equipment Parameter List of Integrated Subsystem

No. :

System (engineering) name						Name of integrated subsystem					
Project manager of integrated subsystem			Contact method			Technical principal of integrated subsystem			Contact method		
Equipment parameter list											
No.	Equipment name	Equipment type	Equipment address	Controllable or not	Equipment explanation	Equipment parameter list					Remarks
						No.	Parameter name	Types of parameter values	Scope of parameter values	Unit of parameter values	Read only	Instruction	
						No.	Parameter name	Types of parameter values	Scope of parameter values	Unit of parameter values	Read only	Instructions	
						No.	Parameter name	Types of parameter values	Scope of parameter values	Unit of parameter values	Read only	Instructions	
						No.	Parameter name	Types of parameter values	Scope of parameter values	Unit of parameter values	Read only	Instructions	
Instruction of corresponding relationship of equipment address											

Table B.0.24 Communication Interface List of Integrated Subsystem

No. :

System (engineering) name		Name of integrated subsystem	
Project manager of integrated subsystem		Contact method	
Technical principal of integrated subsystem		Contact method	
Communication interface types of integrated subsystem		□OPC Data access interface □ODBC Database access interface □Analog video interface □Digital video interface □Data communication protocol (□RS-232 □RS-485 □RS-422) □Ethernet communication protocol(□TCP □UDP)(Adopt the integrated subsystem as: □Server □Client) □Other: _____
Whether the communication interface of integrated subsystem requires to be supplemented with interface equipment or interface software	colspan="3"	□Not required □Require: (successively list the equipment and software to be supplemented) 1. 2.

Submitting accessories description					
No.	Type of accessories	Name of accessories	Instruction of accessories content	Remarks

Construction organization instruction for the integrated subsystem	Principal signs:	Date:
Sign in opinions of the construction organization of integration system	Principal signs:	Date:
Employer organization comments	Principal signs:	Date:
Supervision organization comments	Principal signs:	Date:

Notes:

1. Type of accessories includes: (1) Print (2) Fax (3) Electronic document (4) Software.
2. Where the construction organization of integration system thought that the communication interface provided by integrated subsystem fails to meet the construction demands, items failing to meet the requirements and clauses and subclauses of pursuant design documents or specifications shall be listed successively in the column of opinions.

Table B.0.25 Network Planning and Configuration Table of Intelligentized Integration System

No. :

System (engineering) name				Construction organization		
Public website IP address quantity required				Internal website IP address quantity required		
Demand schedule for IP address distribution						
Filled in by the construction organization of the integrated subsystem				Filled in by the development organization or user's organization		Remarks
No.	Equipment category	Equipment usage	Requirements to IP address	IP address distribution result		
				Whether IP address is automated acquisition		
				IP Address		
				Subnet mask		
				Default gateway		
				DNS Server		
				Whether IP address is automated acquisition		
				IP Address		
				Subnet mask		
				Default gateway		
				DNS Server		
				Whether IP address is automated acquisition		
				IP Address		
				Subnet mask		
				Default gateway		
				DNS Server		
Equipment which is able to access network is required						
Other network planning and configuration requirements						

Employer organization	User's organization	Supervision organization	Construction organization
Principal Date: Seal:	Principal Date: Seal:	Principal Date: Seal:	Principal Date: Seal:

Notes:
1 This table shall be accompanied with a portion of intelligentized integration system network topology diagram.
2 Equipment category includes: (1) Server (2) Workstation (3) Embedded equipment (4) Others.

Table B.0.26 Inspection Record Sheet of Electrical Grounding Device

No. :

Engineering name			Model	
Grounding types		Group numbers	Design requirement	

Grounding device plan sketch (the drawing proportion shall be proper, and numbers and relevant dimensions of each group shall be noted)

Inspection table of grounding device laying conditions (dimensional unit: mm)			
Trench dimension		Soil condition	
Grounding specification		Drilling depth	
Specification of grounding electrode		Welding conditions	
Antiseptic treatment		Grounding resistance	(adopt the maximum value) Ω
Inspection conclusion		Inspection date	

Signature column	Employer (supervision) organization	Construction organization		
		Professional technical director	Professional quality inspector	Professional headman
	Date:	Date:	Date:	Date:

Note: This table is filled in by the construction organization, and the employer organization, construction organization and urban construction archives respectively hold one copy of this table.

Table B.0.27 Grounding Resistance Test Record

No.:

Engineering name				Test date		
Instrument model		Weather condition			Air temperature (℃)	
Grounding types	☐Lightning grounding ☐Machine room grounding ☐Working grounding ☐Protective earthing ☐Static electricity protective grounding ☐Logic grounding ☐Iterative ground ☐Integrated grounding ☐_____					
Design requirement	☐≤10Ω ☐≤4Ω ☐≤1Ω ☐≤0.1Ω ☐≤____Ω					
Test conclusion:						
Employer (supervision) organization	Construction organization					
	Professional technical director		Professional quality inspector		Professional test	
Date:	Date:		Date:		Date:	

Note: This table is filled in by the construction organization, and the employer organization, construction organization and urban construction archives respectively hold one copy of this table.

Explanations of Wording in This Code

1 Words used for different degrees of strictness are explained as follows in order to mark the differences in executing the requirements in this code:

    1) Words denoting a very strict or mandatory requirement:

        "Must" is used for affirmation, "must not" for negation;

    2) Words denoting a strict requirement under normal conditions:

        "Shall" is used for affirmation; "shall not" for negation;

    3) Words denoting a permission of a slight choice or an indication of the most suitable choice when conditions permit:

        "Should" is used for affirmation; "should not" for negation;

    4) "May" is used to express the option available, sometimes with the conditional permit.

2 "Shall meet the requirements of…" or "shall comply with…" is used in this code to indicate that it is necessary to comply with the requirements stipulated in other relative standards and codes.

List of Quoted Standards

GB 50034　*Standard for Lighting Design of Buildings*

GB 50057　*Design Code for Protection of Structures against Lightning*

GB 50093　*Code for construction and acceptance of automation instrumentation engineering*

GB 50116　*Code for Design of Automatic Fire Alarm System*

GB 50166　*Code for Installation and Acceptance of Fire Alarm System*

GB 50169　*Code for Construction and Acceptance of Grounding Connection Electric Equipment Installation Engineering*

GB 50174　*Code for Design of Electronic Information System Room*

GB 50194　*Safety Code of Power Supply and Consumption for Installation Construction Engineering*

GB 50198　*Technical Code for Project of Civil Closed Circuit Monitoring TV System*

GB 50200　*Technical Code for Regulation of CATV System*

GB 50236　*Code for Construction and Acceptance of Field Equipment, Industrial Pipe Welding Engineering*

GB 50263　*Code for Installation and Acceptance of Gas Extinguishing Systems*

GB 50300　*Unified Standard for Constructional Quality Acceptance of Building Engineering*

GB 50303　*Code of Acceptance of Construction Quality of Electrical Installation in Building*

GB 50311　*Code for Engineering Design of Generic Cabling System*

GB 50312　*Code for Engineering Acceptance of Generic Cabling System*

GB/T 50314　*Standard for Design of Intelligent Building*

GB/T 50326　*The Code of Construction Project Management*

GB 50339　*Code for Acceptance of Quality of Intelligent Building Systems*

GB 50343　*Technical Code for Protection Against Lightning of Building Electronic Information System*

GB 50348　*Technical Code for Engineering of Security and Protection System*

GB/T 50356　*Code for Architectural Acoustical Design of Theater Cinema and Multi-use Auditorium*

GB 50371　*Code for Sound Reinforcement System Design of Auditorium*

GB/T 50375　*Evaluating Standard for Excellent Quality of Building Engineering*

GB 50394　*Code of Design for Intrusion Alarm Systems Engineering*

GB 50395　*Code of Design for Video Monitoring System*

GB 50396　*Code of Design for Access Control Systems Engineering*

GB 50462　*Code for Construction and Acceptance of Electronic Information System Room*

GB/T 134　*Methods for Picture and Audio Subjective Assessment of Digital Television Receiving Equipment*

GB 16806　*Automatic Control System for Fire Protection*

JGJ 16-2008　*Code for Electrical Design of Civil Buildings*

JGJ 46　*Technical Code for Safety of Temporary Electrification on Construction Site*

JGJ 146　*Standard of Environment and Sanitation of Construction Site*

GY 5078　*Technical Code for Safety of CATV Distributed Networks Engineering*

GY/T 106　*Technical Specification of CATV Broadcasting System*

YD 5017　*Technical Specification of Execution and Acceptance of Equipment Installation Engineering for Satellite Communication Earth Station*

YD/T 5076　*Planning Specifications for SPC Exchange Installation Engineering*

GY 5073　*Code for Construction and Acceptance of CATV Network Engineering*

GA/T 269　*Black-white Video Doorphone System*

GA/T 72　*General Specifications of Building Intercom System and Elec-control Anti-burglary Door*

GY/T 180　*Technical Specifications of HFC Network Physical Upstream Transmission Path*

GYJ 40　*Acceptance and Test Specifications for Satellite Television Earth Receiving Stations*

YD/T 5120　*Specification on Indoor Coverage Engineering Design For Wireless Communication System*

YDJ 31　*Technical Specifications for Installation Engineering Construction and Acceptance of Communication Power Supply Equipment*

BMB 17　*Technical Requirements for the Gradational Protection of Information System Involving State Secrets*

YD/T 5028　*Temporary Regulations of Engineering Design for Very Small Earth Station VSAT System of Domestic Satellite Communication*

YD/T 5050　*Code for Engineering Design of the Domestic Satellite Communication Earth Station*

YD 5077　*Accepting Specifications for SPC Exchange Installation Engineering*

GY/T 221　*Specifications and Methods of Measurement of Digital Cable Television System*

GY/T 149　*Methods of Measurement for Satellite Digital Television Receive-only Earth Station - System Measurement*

GY/T 151　*Methods of Measurement for Satellite Digital Television Receive-only Earth Station -Outdoor Unit Measurement*